GENESIS AND UTILISATION OF WASTE LANDS

GENESIS AND UTILISATION OF WASTE LANDS

A Case Study of Sultanpur District

HRIDAI RAM YADAV

CONCEPT PUBLISHING COMPANY, NEW DELHI

The Indian Council of Social Science Research gave the financial support towards the publication of this work. However, the responsibility for the contents of book—facts, opinions, conclusions, etc.,—is entirely that of the author and not of the ICSSR.

First Published 1986

Haridai Ram Yadav (*b.* 1957-)

Published by

Naurang Rai
Concept Publishing Company
H-13, Bali Nagar
NEW DELHI-110015 (India)

Printed in India by

Lohia Composing Agency
at Sunil Printers
CB—1067, 75/1 Ajay Palace, Naraina,
NEW DELHI-110028

Dedicated
To
My Mentor
Prof. Moonis Raza

ACKNOWLEDGEMENTS

Words fail to express my thanks to Prof. Moonis Raza for his supervision during the course of this work. His suggestions at every stage of this study and constructive criticisms have helped me immensely in developing this thesis. His humanitarian outlook and constant encouragement have left an everlasting impression on the canvas of my mind.

My greater debt is due to Dr. S. Padmaja who has supervised this work with great patience and helped me with suggestions after going through all the drafts of this work within the shortest possible time.

I am also grateful to Prof. M.K. Premi, Chairman, Centre for the Study of Regional Development, JNU, New Delhi, for allowing me to use various research facilities, providing financial assistance and continuous encouragement.

My heartiest thanks are due to Prof. G.S. Bhalla, Prof. A. Ahmad and Dr. M.H. Qureshi for their inspiration, suggestions and comments at various stages of this work.

Thanks are also due to Shri K. Chandra Mauli, District Magistrate, Sultanpur, Shri Ram Raj Singh Yadav, Sub-Divisional Magistrate, Kadipur, Shri V.P. Pandey, Deputy Director, Soil Conservation, Sharada Sahayak Command, Sultanpur, Kamala Nehru Farm Science Centre, Sultanpur, and other officials of the District for their constant help at the time of field work in collecting required information within a very limited time.

Shri Murlidhar deserves special thanks for his assistance and cooperation in running the computer programme and helping me in systematising the results.

I am no less thankful to my friend Shri M. Ishtiyaqe,

Shri P.N. Tyagi for providing cartographic equipments, Shri S.D.S. Rawat for preparing the typed manuscript, Shri H. Bose for zeroxing the maps and Shri Chinmay Nath for photocopying the maps.

Last but not least, my respectful thanks to my parents and gratitude for brother—B.R. Yadav.

New Delhi
12 November 1984

HRIDAI RAM YADAV

CONTENTS

LIST OF MAPS

LIST OF TABLES

1

INTRODUCTION

1.1 THE PROBLEMS

Agriculture is the mainstay of Indian economy and in India it is not merely an occupation or a business proposition for the people, but is a tradition, a way of life which for centuries has shaped their thoughts, outlook and culture. There is probably no country in the world where land is expected to do so much as in India. Ariculture dominates the country's economy to such an extent that a population of 24.71 crores (37.55% of the total population and of this as much as 66.52% of its working population) is dependent on agriculture.[1]

Importance of agriculture arises mainly because it is the most important sector of the Indian economy contributing nearly 49.6% of the national income. It has been the source of raw materials to India's leading industries and provides a large proportion of her traditional exports. This sector of primary activity being the back-bone of the national economic scene, improvement in the agricultural standards can greatly contribute to the prosperity of the entire Indian economy.

India's agricultural statistics reveal that out of the total area, 46.89% land is effectively under cultivation whereas 16.79% of the area can be categorised as waste lands leaving the rest for non-agricultural uses. There is 19.78% growth in foodgrain production from 1970-71 to 1980-81 and 24.78% population growth during the same decade. Relating the available cultivable land and pressure of population, the

1. *Census of India*, 1981, Series I—India Part IIB (i) Primary Census Abstract, General Population Table.

demand for foodgrains is very high because the population growth has outstripped the rate of increase in agricultural production. On the other hand India's pressure of population and demand for available cultivated land is not corresponding with each other. The per capita cultivated land, which has a value of 0.27 in India, is, by any standard, a lower one. This limited available cultivable land has to meet the demand of the rapidly increasing population, whereby an increase in the agricultural production becomes inevitable. This can be achieved by adopting most suitable strategies as:

(i) Increasing the area under cultivation by putting the present additional marginal uncultivated patches to proper use.
(ii) Increasing the output by multiple cropping.
(iii) Increasing the per hectare yield, because in the last analysis there is no doubt that this is the only way to achieve the requisite agricultural production level in the country.

Assessing the relative merits of each strategy, which are becoming increasingly convincing, the latter two methods are in wide public use. It is a known fact that all efforts are being made to increase the per hectare yield by adopting the most suitable technological means but this alone may not fulfil the demand for foodgrains. Thus it is essential to bring under plough as much area as possible, not only for increasing production but also for providing land to the landless or for improving the size of the existing land holdings. It is for this reason that the cultivation of marginal lands for agricultural purposes has become very essential. Marginal lands include the waste lands, the term used in the latter part of this text. But this has to be assessed in two ways, whether the task is manageable physically as well as economically. The Planning Commission is of the view that this is a feasible step keeping in view the cost incurred in bringing the waste lands into cultivation. The recent experiments in the north Indian belt have provided enough data to explore the possibility of bringing temporary waste lands back into cultivation.

The marginal uncultivated patches include those types of

areas which are cultivable or in fact were cultivated some time back but at present are lying uncultivated due to some reason or the other. Such lands may also be termed as "Waste Lands". Bringing such lands under cultivation may provide ample scope for increasing agricultural production. But such an attempt would need proper categorisation of waste lands and their major determinants. Once this is done, it would become easier to find out the ways of bringing them back under cultivation.

1.2 LITERATURE SURVEY

From the detailed literature survey conducted for this theme of research, it is found that the studies which are directly related to the waste lands are very scant. However a few relevant studies are mentioned below. Dac Greger[2] and Kennedy[3] have explained the distribution of waste lands and have suggested suitable measures for their reclamation. The Ministry of Food and Agriculture[4] has conducted a survey on location of the waste lands distribution in India (Statewise reports) with a few reclamation measures. Hadimani[5] has studied the waste lands distribution and reclamation of the same in Delhi State (now a Union Territory). Shafi[6] has discussed waste lands distribution in India with particular reference to other uncultivated land excluding fallows and fallows other than current fallows, saline-alkali soils, water-logged land and ravine land of India. In this study he has analysed the reclamation measures of the above waste lands which were carried out in different parts of India by the Waste Lands

2. Dac Greger, N.M. & Datta, S.K., "The Utilization of Waste Lands", 1947, *Indian Tea Association Tockali*, Exp. Sta. Mem. No. 20, Calcutta, p. 19.
3. Kennedy, K., "Australia's waste land is now bountiful", 1962, *Indian Farming II*, (10), p. 29.
4. Waste Lands Survey and Reclamation Committee, Ministry of Food and Agriculture, 1961, *Report on Location and Utilization of Waste Lands in India*, Part X. Uttar Pradesh.
5. Hadimani, A.S., "Studies on Waste Lands of Delhi State", 1960, *Msc. Associate, I.A.R.I.*, New Delhi.
6. Shafi, M., "The Problem of Waste Lands in India" Published in *Geographer*, Special Number XXI, Aligarh Muslim University Geographical Society, Aligarh, 1968.

Survey and Reclamation Committee. He has also made a cost-benefit analysis of the waste lands. The work done, by Abha Lakshmi Singh,[7] on the Economics and Geography of Agricultural Land Reclamation in Koil Tahsil of Aligarh District, in which she had taken all 358 villages of the tahsil, is a new line of approach to waste land studies. In the first phase, a survey of land, classified as other uncultivated land excluding fallows and fallows other than current fallows was conducted to locate the areas where large blocks of land are available for reclamation and resettlement. Each and every village map was collected from the tahsil headquarters at Aligarh. The general aspects including the relief, drainage, climate and soil of the area are also studied in detail. Suitable measures for the reclamation, according to the conditions in different areas, estimate of the cost of reclamation and colonisation, the financial assistance and agricultural extension and the necessary training for the settlers are suggested. Some of the *usar* prominent blocks have also been studied. Finally the author has estimated the economic aspects of such reclamation in terms of the expenditure involved, likely addition to the foodgrain production, employment and the income that would become available to the settlers.

The literature reviewed for this work proves that most of the studies are concerned with the problems and possible reclamation of saline and alkaline soils. United States of America Salinity Lab[8], Academy of the Sciences of USSR[9] and Gedroiz[10] have done a detailed analysis of the saline land distribution and its reclamation. In the reclamation measures the USSR Academy has discussed the application of drainage

7. Singh, Abha Lakshmi, *Economic and Geography of Agricultural Land Reclamation*, 1978, B.R. Publishing Corporation, Vivekanand Nagar, New Delhi.
8. U.S.A. Salinity Lab., "Diagnosis and Improvement of Saline and Alkali Soils—Handbook", *U.S. Department of Agriculture*, 1954, No. 60.
9. Academy of the Sciences of USSR, "The Application of Drainage in the Reclamation of Saline Soils, 1958, translated from Russian, Moscow.
10. Gedroiz, K.K., "Saline Soils and their Improvement", *Journal of Experimental Agronogy*, 18 (in Russian, translated by Waksman, S.A.), 1977.

and other scientific reclamation measures for the salinized soils. The work done by Uppal[11], Kanwar[12], Yadav and Aggarwal[13], Nigam[14] provides the measure to reclaim saline and alkali soils along with the assessment of their potential. Their economic aspects of utilisation are also provided by them, adopting various reclamation measures, viz. leaching, through irrigation and other scientific methods, and they have also discussed saline soils reclamation through dhaincha and gypsum application. Barnes[15], Dhar[16], Uppal[17], Aggarwal[18] and Singh[19] have worked on the reclamation of Alkali waste lands through biochemical factors while Prashad and Yadav[20] have prepared a list of questions and answers for the alkali and saline soil reclamation. They have also pointed to water quality, drainage measures, cultivation (crop and varieties) sowing time, nutrient requirements, control of pests and diseases, economics of reclamation, facilities for reclamation and required educational

11. Uppal, H.L., "Reclamation of Saline and Alkali Soils in India", *Directorate of Extension Farm Bulletin*, No. 66, New Delhi, 1961, p. 30.
12. Kanwar, J.S., "Reclamation of Saline, Alkali Soils, Amendments, Kinds, Amounts and Costs", paper presented in the *Symposium on Salinity and Alkalinity, I.A.R.I.*, New Delhi, 1962.
13. Yadav, J.S.P. & Aggarwal, R.R., "Dynamics of Soil changes in the Reclamation of Saline, Alkali Soils of the Indo-Gangetic Alluvium", *Journal of Indian Society, Soil Science*, 7, 1959.
14. Nigam, N.M., "How to Reclaim Saline and Alkali Lands in Delhi State", Thesis Associate, I.A.R.I., New Delhi, 1956.
15. Barnes, J.H. and Barkat Ali, "Alkali Soils, Some Bio-chemical Factors in their Reclamation", Report on Bihar and Orissa, Department of Agriculture, 1930-31, *Agricultural Journal India*-12.
16. Dhar, N.R., Mukherjee, S.K. and Biswas, N.N., "Alkali Soils and their Reclamation", *Current Reports, Academy of Agriculture*-23, 1938.
17. Uppal, H.L., "Reclaming Alkali Waste Lands", *I.C.A.R., Bulletin* No. 85, 1961, New Delhi, p. 40.
18. Aggarwal, R.R., "Delhi Soils can be Reclaimed", *Indian Farming*, 7 (9), 1957.
19. Singh, N.T., "Reclamation of Alkali (Kallar) Soils", 1972, Punjabi Agricultural University, Ludhiana, p. 16.
20. Prashad, R. and Yadav, J.S.P., "Alkali Soil Reclamation", Questions and Answers, Central Soil Research Institute, Karnal, India, 1969.

training for the reclamation by farmers. Leather[21], Khan[22], Khan[23], Kaul[24], National Botanic Gardens in Lucknow[25], Reh Committee[26], United Provinces of Agra and Oudh and the Usar Land Reclamation Committee[27] have discussed the usar land identification, their diagnosis—and investigations into the causes of deteriorating soil fertility by Reh in different areas. They have analysed reclamation of usar areas with suitable measures through irrigation, drainage, gypsum and pyrites and other scientific methods. Aggarwal[28], Auden[29], Milne[30], Raheja[31] and Hill's[32] work is also on the same pattern with

21. Leather, J.W., "Reclamation of Reh or Usar Soils", *Agriculture Ledger*, 1893, pp. 12-13.
22. Khan, A.V., "Diagnosis and Reclamation of Usar Soils", *Technical Bulletin* No. 4, Lucknow, March 1951.
23. Khan, A.D., "Usar Soils, their Identification and Reclamation", *Agriculture Animal Husbandry*, U.P., (5), 1950.
24. Kaul, K.N., "Scheme for Reclamation and Development of Usar Lands", Report, I.C.A.R. Scheme—1961.
25. National Botanic Gardens, Lucknow, "Conquest of Usar", 1961 (An experiment on reclamation of Alkali land at Banthara, Lucknow District, by Biological and Agronomic Methods).
26. Reh Committee, "Report of the Committee for Investigating into the Causes of the Deterioration of land by Reh in Aligarh District, in Alkali lands, Irrigation and Drainage, 1878". By Hilgard, E.W., *University of California*, Appendix XVII to Report for the year 1886 and also in *Dictionary of Economic Products*. Vol. XI. Part I.
27. United Provinces of Agra and Oudh, Department of Agriculture, 1909, "Reclamation of Usar Land, Report of Agra and Oudh" U.P. 1908-1909.
28. Aggarwal, R.R. and Yadav, J.S.P., "Saline and Alkali Soils of Indo-Gangetic Alluvium in U.P.", 1954, *Journal of Soil Science*, 5 (2).
29. Auden, J.A. and others, "Report on Sodium Salts in Reh Soils in the United Provinces with notes on occurrences in other parts of India, Records, Geological Survey of India 1942-77 (I).
30. Milne, R.A. and Rapp, E., "Soil Salinity and Drainage Problems, Causes, Effects and Management, U.S. Department of Agriculture, 1968, *Farmers Bulletin*, No. 1314, Washington, p. 26.
31. Raheja, R.C., "Study of Suitable Trees, Grasses and Crops for Saline and Alkali Land", 1962, prepared by the Working Group on Waste Land including Saline, Alkali and Waterlogged Land, Government of India.
32. Hill, E.G., "The Analysis of Reh, The Alkaline Salts in Indian Usar Lands, 1903, *Chemical News*, p. 87.

stress on saline and alkali land, waterlogged land and their distribution and characteristics along with which they have suggested suitable reclamation measures through proper scientific management in accordance with their problems, causes and effects. Even the study of Raheja provides many valuable suggestions for the reclamation of saline and alkali soils and recommends planting of trees, growing of grasses, and crops, and the other scientific methods. Pilai[33], Raj[34], Kamps[35] and the Committee on Natural Resources, Planning Commission[36], have studied in detail the waste lands where they have also analysed the distribution and reclamation of alkali and waterlogged lands in different parts of the country basing their case studies in Kerala and other States. Ray Chaudhury[37] has discussed the extension of cultivated areas for increasing agricultural production in connection with the waste lands. Ray Chaudhury,[38] Indian Council of Agricultural Research (ICAR)[39] and State Soil Conservation Research[40] have carried out a survey of soils of India and have suggested soil conservation measures through a scientific reclamation of waste lands. Taylor[41], Kanwar[42],

33. Pilai, V.R. and Panikar, P.G.K., *Land Reclamation in Kerala* Asia Publishing House, 1965, New Delhi.
34. Raj, S.K.T., "Ganga Khadir, A Study of Land Reclamation in India", Ministry of Information and Broadcasting, 1951, *Pamphlet* No. 4, New Delhi, p. 44.
35. Kamps, L.F., "Mud Distribution and Land Reclamation in Eastern Wadden Shallows", Netherlands International Institute for Land Reclamation and Improvement, Wageningbn.
36. India's Planning Commission : Natural Resource Committee, New Delhi, 1963, "Study on Waste Lands including Saline, Alkali, Waterlogged Land and their Reclamation".
37. Ray Chaudhury, S.P., "Extension of Cultivation Area for increased Agricultural Production", *Proceedings of the Symposium on Science and India's Food Problem*, I.C.A.R. Publication, 1977, p. 161.
38. Ray Chaudhury, S.P., "Final Report of the All India Soil Survey Scheme, *Bulletin—ICAR*, 1952 (13).
39. I.C.A.R., Final Report of All India Soil Survey Scheme, 1958.
40. State Soil Conservation Research, "Annual Report, State Soil Conservation Research U.P.", 1961.
41. Taylor, E.M., "Making Land Reclamation Profitable, *Indian Farming*, 1940, Vol. I (9).
42. Kanwar, J.S., "Recliaming Deteriorated Soils", *Fertilizer News*, 6, 1961 (21).

Bhattacharya[43] and the Planning Commission, Government of India, have studied the distribution of waste lands and suggested suitable measures for soil conservation, to make the soil more fertile.

Bali[44], Sharma[45], Ahmad[46], Sharma[47] and the Natural Resources Committee, Planning Commission[48], have surveyed the ravine distribution in India. Sharma has studied, in special, the genesis and distribution of ravines of the Chambal Valley. He considers rejuvenation as the main cause for the ravine formation and suggests a few measures for their reclamation.

It is generally seen that most of these studies are concentrated on the problems and reclamation of mainly saline and alkaline soils and waterlogged lands. On the other hand a few studies conducted by geographers are mainly concerned with spatial distribution of waste lands of a particular area, of a particular type and at a single point of time. Primarily there has also been an attempt by them to study the possible reclamation measures along with its economics. Secondly, it can also be observed that the main thrust is on the pedogenic factor, while probing into the causes of waste land formation.

There are shortcomings in the previous studies which mainly are :

(i) That though they are problem oriented, they lack adequate explanation, where the thrust has been only on a few determinants.

(ii) As they are carried out by the Government agencies, the picture of the totality is lost, in the sense that they are limited to the extent of the chosen problem.

43. Bhattacharya, J.P., "Study of Soil Conservation for Agricultural Land", 1964, New Delhi.
44. Bali J.S., "Ravine Reclamation needs for India", 1972, New Delhi.
45. Sharma, H.S., *Ravine Erosion in India*, 1980, Concept Publishing Co., New Delhi.
46. Ahmad, E., *Soil Erosion in India*, 1970, Asia Publishing House, New Delhi
47. Sharma, H.S., "Genesis of Ravines of the Lower Chambal Valley, India", 1968. *Selected Papers*, 21st International Geographical Union Congress, 1, pp. 114-18.
48. India, Planning Commission : Natural Resource Committee, "Study on Survey and Reclamation of Ravines in India", *1965*, New Delhi.

(iii) Though the reclamation measures for the waste lands are suggested but the capability classification associated with their potential is lacking.

Scope for modifications for the present study exists in :

(i) Identifying the typology of waste lands at all the levels in a spatial and temporal context which would lead to a greater comparative analysis. Such an analysis would also help in identifying the relative merits and demerits of waste land reclamation in general.
(ii) Assessing the relative merits of waste lands.
(iii) Finding the determinants of waste land formation.
(iv) Identifying the capability zones of waste lands.
(v) Analysing the genesis of waste lands and remedies for their reclamation.
(vi) Estimating the economics of reclamation.

1.3 OBJECTIVES OF STUDY

On the basis of the scope for the modifications, the present study aims at the following objectives:

(1) The main aim is to categorise the waste lands at various areal levels, i.e. tahsil, block and village, of Sultanpur District and to explain the growth of waste lands from 1977-78 to 1981-82;
(2) To identify various environmental and anthropogenic factors which may be responsible for the formation of classified waste land types;
(3) To find out various correlates of waste lands;
(4) To explain the genesis of waste lands through geomorphic processes involved in the development of waste lands in association with various factors;
(5) To explain the limitations of each type of waste land;
(6) To suggest suitable remedies for the reclamation of each type of waste land;
(7) To assess the economics of waste land reclamation through an input-output balance analysis.

1.4 CHOICE OF THE INDICATORS

The indicators chosen for the present study are broadly grouped into two categories of dependent and independent variables. The choice of the independent variables from numerous parameters needs consideration in the sense that they should, in every possible way, show an association with the chosen dependent variables. Henceforth these variables can be of various types representing both the natural and cultural arenas.

1.4(A) Dependent Variables (Waste Lands)

(1) Water Logged Land, (2) Usar Land, (3) Banjar Land, (4) Old Fallow Land, (5) Fallow Land, (6) Kankrili Land, (7) Ravine Land, (8) Other Types of Waste Land.

1.4(B) Independent Variables

(a) Morphometric Factors: (1) Slope, (2) Ruggedness Number.

(b) Climatic Factor: (3) Mean Annual Rainfall.

(c) Hydrologic Factors: (4) Drainage Density, (5) Quality of Water in *p*H, (6) Behaviour of Water Table.

(d) Pedogenic Factors: (7) Nitrogen of the Soil, (8) Phosphate of the Soil, (9) Potash of the Soil, (10) *p*H Value of the Soil, (11) Thickness of the Top Soil, (12) Depth of the Base Rock.

(e) Anthropogenic Factors: (13) Distance of the Canal from the Field, (14) Land Concentration (Gini's Coefficient).

1.5 HYPOTHESIS

(1) Environmental and anthropogenic parameters (independent) show varying trends of inter-relations with various types of waste land (dependent variable).

(a) It has been assumed that waterlogged land may have a positive correlation with slope, ruggedness number, mean annual rainfall, drainage density and potash of the soil and it may have a negative correlation with quality of water in

pH, behaviour of water table, nitrogen, phosphate, pH value of the soil and distance of the canal from the field.

(b) Assumption has been made that usar land may have a positive relation with the quality of water in pH, nitrogen, phosphate, potash and pH value of the soil and a negative relation with slope, ruggedness number, mean annual rainfall, drainage density, behaviour of water table and thickness of the top soil.

(c) Banjar land may have positive correlation with slope, ruggedness number, mean annual rainfall, drainage density, quality of water in pH, behaviour of water table, potash and pH value of the soil, land concentration and a negative relation with the nitrogen and phosphate of the soil.

(d) Old fallow land may have a positive relationship with slope, ruggedness number, drainage density, quality of water in pH, behaviour of water table, pH value of the soil and land concentration and it may have a negative correlation with the mean annual rainfall, nitrogen, phosphate, potash of the soil.

(e) It is assumed that fallow land may have a positive relationship with slope, ruggedness number, mean annual rainfall, drainage density, quality of water in pH, behaviour of water table, pH value of the soil, land concentration and negative relationship with nitrogen, phosophate and potash of the soil.

(f) Assumption has been made that *kankrili* land may have a positive correlation with slope, ruggedness number, mean annual rainfall, drainage, density, quality of water in pH, behaviour of water table, potash, pH value of the soil and depth of the base rock and negative correlation with the nitrogen and phosphate of the soil.

(g) Ravine land may have a positive correlation with slope, ruggedness number, drainage density, quality of water in pH, behaviour of water table, pH value of the soil and it may have a negative relationship with the mean annual rainfall, nitrogen, phosphate and potash of the soil.

(h) It has been assumed that the other types of waste land may have a positive correlation with slope, ruggedness number, mean annual rainfall, drainage density, behaviour of water table, pH value of the soil and land concentration and it may have a negative correlation with the quality of water in pH, nitrogen, phosphate and potash of the soil.

(2) The waste lands are a result of a complex interaction of processes and can be designated as compound land forms.

(3) The intense the stage of development of waste lands, the lower their capability and higher their cost of reclamation.

1.6 DATA BASE

For any type of study there should be a proper dependable data base which can be utilised for the formation of a theory. This data base varies from one theme to another depending on the nature of the problem. In the present study data base can be categorised under the following heads:

(A) Records

(1) Land Revenue Records (Milan Khasara Bandobast) at Muhafijkhana, headquarter of each tahsil and their officials as lekhpals, kanoongos, registrar kanoongos, Naib Tahsildars and Tahsildars of each tahsil have been communicated personally;
(2) Statistical book of the District Sultanpur 1981-82.
(3) Milan Khasara of each tahsils from Sadar Kanoongo for the year 1977-78.
(4) Kharif Abhian by Agriculture Office Sultanpur—1981-82;
(5) Rainfall data is collected from the tahsils and block headquarters of the District Sultanpur:

(B) Field Work

(1) Most of the data has been collected at the time of field work interviewing revenue officials and villagers including village heads and through field observations.

(C) Maps

(1) Survey of Indian Topo-sheets: 63 F/10, 11, 12, 14, 15, 16, 63 J/3, 4, 7, 8, 11, 12, 63 K/5, 9.

(D) Personal Communication

(1) Office of the District Magistrate and Sub-Divisional Magistrates.
(2) Office of the Deputy Director, Soil Conservation, Sharada Sahayak Command, Sultanpur.
(3) Office of the Soil Conservation, Sultanpur District, and Agriculture Officer.
(4) Office of the Executive Engineer, Sharada Sahayak Khand 16 and 49, Sultanpur.
(5) Office of the Executive Engineer (Tubewell), Sultanpur.
(6) Kamala Nehru Farm Science Centre, Sultanpur.
(7) Office of the Director, Social and Forestry Department, Sultanpur.

(E) Laboratory

(1) District Soil Laboratory, Sultanpur.

1.7 METHODOLOGY

The adoption of a suitable methodology provides an incentive to analyse and represent the data in a proper form which in turn helps either to uphold or reject the hypothesis formulated for the study. Once again the choice of the methodology for the purpose is an essential aspect of the study. The methodologies adopted in this study are as follows:

1.7 (A) Quantitative Techniques

1.7 (A) (1) Waste Lands : In Sultanpur District, eight types of waste lands are observed which mainly are waterlogged land, usar land, banjar land, old fallow land, fallow land kankrili land, ravine land and other types of waste land. The data is available only for the usar land, banjar, old fallow, fallow and other types of waste land with the revenue offices of the district but for the waterlogged, kankriii and ravine lands Lekhpals of the villages of the district are contacted personally and the data collected at the time of field work for all the villages. After collection of the data of all waste land types at village, block and tahsil levels of the district for the years 1977-78 and 1981-82, percentage of each waste land type to the total area of the respective units and the percentage growth of each waste land type have been calculated for the same period of time.

1.7 (A) (2) Slope : The topo-sheets have been consulted to find out the heights of the villages. The available spot heights or contour values have been considered and have been plotted on the village map while villages are surveyed properly and according to the local elevations, some more additional spot heights are plotted on the same map and lastly, the contour lines and drawn are from this minimum and maximum contour lines and identified are the average number of contour crossings per mile is computed. The average slope is calculated using the Wentworth formula :

$$\text{Average Slope} = \frac{\text{Average Number of Contour Crossings per mile} \times \text{Contour Interval}}{3361 \text{ (Constant)}}$$

The computed slope has also been crosschecked from the surveyed slope by the office of the Executive Engineer (Irrigation), Sharada Sahayak Khand.

1.7 (A) (3) Ruggedness Number : According to the field observations drainage lines are drawn on the actual map of the village and the total length as such is divided by the area to calculate the drainage density of the village and relative relief has been computed subtracting the minimum height from the maximum height of the village. Lastly ruggedness number is computed :

$$\text{Ruggedness Number} = \frac{\text{Drainage Density} \times \text{Relative Relief}}{5280}$$

1.7 (A) (4) Mean Annual Rainfall : The data for the mean annual rainfall is available at the district and tahsil and a few blocks of the district and to know the mean annual rainfall of the selecled villages, the collected data has been extrapolated for the year 1981-82.

1.7 (A) (5) Drainage Density : It has already been discussed that the natural and artificial drainage lines are drawn for the selected villages and the length of the drainage lines per unit area are measured and, lastly, the drainage density is computed as follows :

$$\text{Drainage Density} = \frac{\text{Total Drainage Length}}{\text{Total Area}}$$

1.7 (A) (6) Quality of Water in *p*H : The information for the quality of water in *p*H is collected from the tubewell irrigation department, office of the Executive Engineer and, in a few cases, the *p*H value of the water has been tested by the same department.

1.7 (A) (7) Behaviour of Water Table : The data for the behaviour of the water table is collected from the office of the Executive Engineer (Tubewells), Nalkup Khand division, and the ground water survey division of the district. The data is also supplemented by personal field observation of the wells at the time of field work, the depth of which is tape-measured and, lastly, the data is generalised at village level.

1.7 (A) (8) Nitrogen, Phosphate, Potash and *p*H Value of the Soil: The data for the nitrogen, phosphate, potash and *p*H value of the soil has been collected from the District Soil Laboratory, Sultanpur, and a few soil samples were given to the Soil Laboratory to know the nature of the soil for the required sample villages of the district.

1.7 (A) (9) Thickness of the Top Soil and Depth of the Base Rocks : Data pertaining to the thickness of the top soil and depth of the base rocks has been collected by digging pits and also from the already digged pits of the area.

1.7 (A) (10) Distance of the Canal from the Field : The data for distance of the canal from the field has been collected from the office of the Executive Engineer, Sharada Sahayak Irrigation Khand 16 and 49, office of the Deputy Director, Soil Conservation, Sharada Sahayak Command, Sultanpur.

1.7 (A) (11) Land Concentration (Gini's Coefficient) : Gini's coefficient is a value to know the inequality of the land (area) and the household which is dealt with the cumulative percentage distribution of the two attributes at different points. In this case Household and Area are the two attributes. The inequality between the two attributes can be known through the graphical presentation, i.e. Lorenz Curve. Gini's coefficient in any such graph may be measured numerically in terms of the ratio of the area under the curve and the line of equal distribution to the area of the triangle formed by the X-axis, the Y-axis and the line of equal distribution. Thus Gini's coefficient can be numerically worked out by the following formula :

$$\text{Gini's Coefficient} = \frac{1}{100 \times 100} \left| \sum_{i=1}^{n} X_i Y_i + 1 \left(\Sigma X_i + 1 Y_i \right) \right|$$

1.7 (A)(12) Choice of the Samples : To explain the factors influencing each waste land type of the district fifteen villages were selected on the basis of quartiling, viz. very high, high, medium, low and very low percentage of different types of waste land to the total area of the village where stratified sampling method is used for the balanced representation to explain the determinants of each waste land type. From these, three villages were selected to analyse the genesis of waste lands.

To explain the remedies for the reclamation of waste lands and economics of waste land reclamation one acre land of one village for each waste land type has been selected on merit, i.e. the village having the highest percentage of a particular type of waste land to the total area of the village.

1.7 (A) (13) Explanatory System : The identification of casual relationship between dependent variables (waterlogged land, usar land, banjar land, old fallow land, fallow land, kankrili land, ravine land and other types of waste land) and independent variables (environmental-morphometric, climatic, hydrologic, pedogenic and anthropogenetic factors) is an essential concern of a scientific investigation. More precise quantitative measurement of the degree and direction of correlation is worked out as suggested and defined by Karl Pearson coefficient of correlation[49] and the correlation matrix is prepared to see

49. Formula for the correlation

$$r = \frac{\Sigma XY - \frac{\Sigma X \Sigma Y}{n}}{\sqrt{(\Sigma X)^{2-} \frac{\Sigma X^{2}-}{n}} \sqrt{\Sigma Y^{2} - \frac{(\Sigma Y)^{2-}}{n}}}$$

Significance test of correlation coefficients, using the test of significance of r, it is possible to infer whether correlation coefficient between the variables will be zero or not. Under the null hypothesis

the correlation of independent variables with the various types of waste lands.

The stepwise regression analysis is done as the final step.[50]

the expression as given below will follow the distribution with (n-2) degrees of freedom

$$t = r\sqrt{\frac{n-2}{1-r^2}}$$

where n is the number of observations used and r is the coefficient of correlation.

50. The form of linear regression analysis between dependent variable Y and an independent variable X is given as
$Y = £ + \beta X + U$
Where the constants are the intercept and the slope of the straight line and U is the error term.
The basic objective of a regression analysis is to estimate the values of £ and β. The estimated least square form of the relationship is given by Johnston*

$$Y = a + bX$$

where

$$b = \frac{\Sigma XY - \frac{\Sigma X \Sigma Y}{n}}{\Sigma X^2 - \frac{(\Sigma X)^2}{n}} \text{ is an estimate of } \beta$$

and $a = \overline{Y} - b\overline{X}$ is an estimate of £
Test of the significance which is used is given below

$$t = \frac{b - o}{-SE(b)}$$

with (n—2) degree of freedom.

where $-SE(b) = \frac{\sigma u}{\sqrt{\Sigma X^2 - \frac{(\Sigma X)^2}{n}}}$

$$\text{and } \sigma u^2 = \frac{\Sigma X^2 \quad \frac{(\Sigma Y)^2}{n} - b\left\{\Sigma XY - \frac{\Sigma X \Sigma Y}{n}\right\}}{n - 2}$$

Another important summary statistic in a regression analysis is the coefficient of determination R^2 which tells us the proportion of variations in Y as explained by X and is given by (variation here means sum of squares of deviation from mean and is also known as sum of squares)

$$R^2 = \frac{\text{Explained S.S.}}{\text{Total S.S.}}$$

Where explanation sum of squares

$$= b\left\{\Sigma XY - \frac{\Sigma X \ \Sigma Y}{n}\right\} \text{and}$$

$$\text{Total sum of squares} — = \Sigma Y^2 - \frac{(\Sigma Y)^2}{n}$$

*Johnston, J., *Economics Methods*, McGraw Hills, 1972, pp. 8-16.

This model is chosen mainly because "wherever multiple regression is needed, it is useful to know as to how the parameters get changed when new variables are added, one by one".[51]

1.7 (B) **Cartographic Techniques**—The cartographic representation is made to depict the data in a suitable graphic form. Firstly, choroplething is used to show the distribution of waste lands at village level. Secondly, Bar Diagram is used to present the growth of each waste land type at Block level and, thirdly, Pie (wheel) Diagram is used to present each waste land type at Block level in Sultanpur District.

1.8 CHOICE OF THE STUDY AREA

For a work of this kind primarily concerned with the identification and explanation of aerial variations, the choice of the region for the analysis is very important. Sultanpur District has been selected for the present study because there is an availability of different types of waste land, viz. waterlogged land, usar land, banjar land, old fallow land, fallow land, kankrili land, ravine land and other types of waste land, and the total waste land is recorded as 13.72% to the total area of the district. There is also a 10.93% growth in the total waste lands of Sultanpur District from 1977-78 to 1981-82 and there is much variation in these waste lands at all the three levels, i.e. Tahsil, Block and Village. The growth of each type of waste land at tahsil and block level is also recorded to be significant excepting in Gauriganj Block where waterlogged land has decreased by 30.76% from 1977-78 to 1981-82. Secondly, the variation in the levels of socio-economic development at the three levels leads to a better choice of indicators in explaining the waste land formation.

Lastly, as the area has a potential for agricultural development insurgence of suitable reclamation measures will serve as

51. Mahmood Aslam, *Statistical Methods in Geographical Studies*, Rajesh Publication, New Delhi, 1977, pp. 34-61.

an input into the economic betterment of the region as a whole.

1.9 CHOICE OF THE TIME PERIOD

The choice of the time period is very much a matter of convenience—as the analysis is based on the field observation, the period of field work 1981-82 has been taken as the base year. For the temporal variation a five-year gap is chosen to witness any change in the waste land extension from 1977-78 to 1981-82.

1.10 GEOGRAPHIC BACKGROUND OF THE STUDY AREA

Sultanpur District in Faizabad Division of Eastern Uttar Pradesh is situated between the parallels of 25°59′ and 26°40′ North and 81°32′ and 82°41′ East. According to the Surveyor-General of India the area of the district is 442 Sq. km. Geographically Sultanpur lies in the eastern region of the State. The district is bounded in the east by Jaunpur and Azamgarh Districts, in the west by Rae Bareli and Barabanki Districts, in the north by Faizabad District and in the south by Pratapgarh District. It comprises four tahsils : Sadar Sultanpur (6 Blocks), Kadipur (5 Blocks), Musafirkhana (4 Blocks) and Amethi (4 Blocks). (Map 1). Thus there are 19 Blocks in Sultanpur District and it comprises 2,535 villages (Map 2). The headquarters of Sultanpur District are situated on the right bank of river Gomati.

1.10.1 History of the District

Tradition tells that a town was founded on the left bank of river Gomati by Kusa, son of Rama, and was called Kushbhavanpur. At the time of the Mohammedan conquest, the district was held by the Bhars, but no place of importance was situated within it and no reference to it can be found in the Persian history. Local tradition asserts that Kushbhavanpur was conquered by Ala-ud-din, but the name of the conqueror is probably a mistake. The district formed part of the Jaunpur

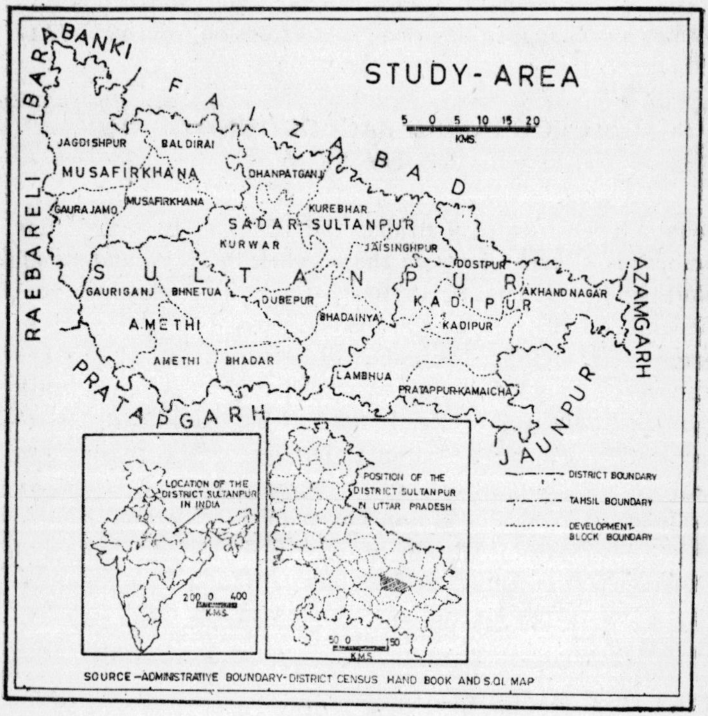
STUDY-AREA
5 0 5 10 15 20
KMS.
BARABANKI
FAIZABAD
RAEBARELI
AZAMGARH
JAUNPUR
PRATAPGARH
SULTANPUR
JAGDISHPUR
BALDIRAI
MUSAFIRKHANA
DHANPATGANJ
GAURA JAMO
MUSAFIRKHANA
KUREBHAR
SADAR-SULTANPUR
KURWAR
JAISINGHPUR
DOSTPUR
GAURIGANJ
BHNETUA
DUBEPUR
KADIPUR
AKHAND NAGAR
BHADAINYA
KADIPUR
AMETHI
AMETHI
BHADAR
LAMBHUA
PRATAPPUR-KAMAICHA
LOCATION OF THE DISTRICT SULTANPUR IN INDIA
200 0 400
K.M.S.
POSITION OF THE DISTRICT SULTANPUR IN UTTAR PRADESH
50 0 150
K.M.S.
DISTRICT BOUNDARY
TAHSIL BOUNDARY
DEVELOPMENT-BLOCK BOUNDARY
SOURCE – ADMINISTRATIVE BOUNDARY- DISTRICT CENSUS HAND BOOK AND S.O.I. MAP

MAP 1

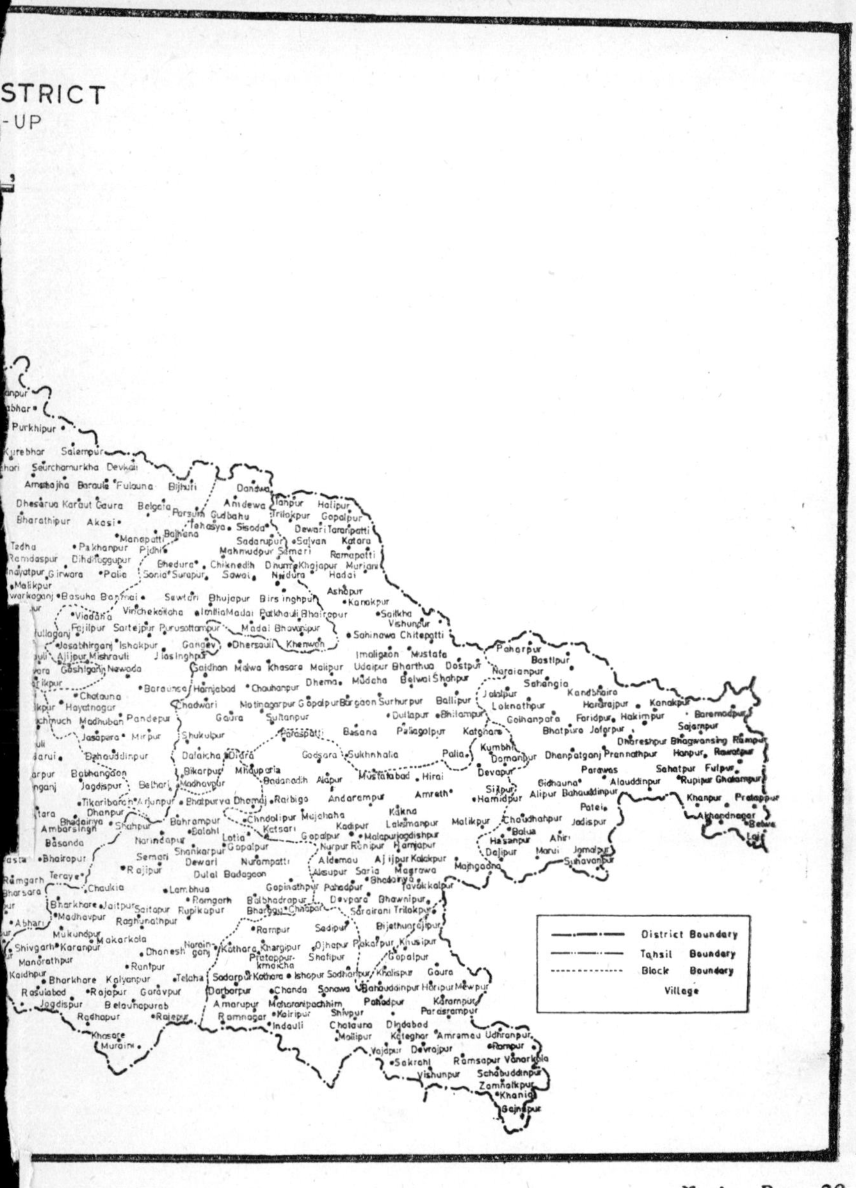

Facing Page 20

MAP 2

kingdom in the 15th century and on the downfall of the Lodi Dynasty was incorporated in the Delhi empire. Under the redistribution made by Akbar the present area fell partly in the Subah of Oudh and partly in that of Allahabad but 250 years later the whole district came under the Nawab of Oudh. In 1856 when Oudh was annexed, a district was formed which included portions of what are now Barabanki and Raebareli districts, while additions have been made to it from Faizabad. The district was named Sultanpur and assumed its present shape in 1869.[52]

Within the district many ancient mounds are found which are connected by local tradition with the Bhars. Some of them have yielded Buddhist remains but no regular excavations have been made. The chief sacred places connected with the story of the Ramayana are Sitakund, a bathing ghat on the Gomati close to Sultanpur and Dhopap lower down the same river. At the latter place are the ruins of a fort built by Sher Shah which is known as Shahgarh.

1.10.2 Geology

There is no noteworthy feature in the physiognomy of the district, or it can be stated that geologically Sultanpur District is not important. The plain is characterised by an unrelieved monotany of alluvial deposits which completely shroud the old land surface to a depth of several thousand meters. The solid geology of the plain is totally obscured underneath the mantle.[53] The main dominating features of the district are the ravines of the Gomati, where only sand is seen. Kankar is also found in the northern portion of the district where some usar lands are also observed. The soils are mainly sandy, loamy and clayey, sub-divided into Khader or newer alluvium. The whole plain of the district is formed of rich alluvial soil which is of great agricultural importance. But the kankar or calcareous limestone is common.

52. *The Imperial Gazettee of India*, Vol. XXIII, New Delhi, 1974, pp. 130-37.
53. Singh, M., "Land Utilization in Eastern Uttar Pradesh", Ph.D. Thesis, unpublished, p. 19.

1.10.3 Physiography

With the exception of a gradual and scarcely perceptible slope from north-west to south-east, the surface of the district is generally level, being broken only by the ravines in the neighbourhood of the river by which its drainage is affected.

The average height of the district is 120 metres. Heterogeneity in the physical landscape is produced actually by local eminences such as river levees and bluffs or sandy features like bhurs, oxbow lakes, tals, chaurs, dead arms or remnants of the river channels or occasionally, available bad lands and ravines. These are the perceptible notches of the banjar tracts. In a real sense the scenery is of a varied character. Many spots along the Gomati are exceedingly pretty, but for the most part on both banks of the river is a dreary, bleak and ravine cut tract, occasionally relieved by sarpats (grass) and mango groves.

The Ganga-Yamuna doab is part of the Indo-Gangetic Plain. This plain is a depression between the Himalayas in the north and the Deccan plateau in the south. It has been filled by the alluvium brought down by the Himalayan rivers. The deposition of this alluvium commenced after the final upheaval of the mountain and has continued all through the Pliestocene up to the present.[54] Various hypothesis have been put forward to explain the geological evolution of this plain. The accurate depth of the alluvium is not known. Boring down mainly for artesian wells has penetrated only upto 1,606 metres in the recent alluvium strata.[55] Geologists differ in their estimates of the thickness of the alluvial deposits. On the basis of the observation Glennie[56] estimated it at about 1,950 metres.

1.10.4 Drainage

Sultanpur district is part of the well-integrated drainage system of the Gomati (Map 3) although different nalas join the master

54. Wadia, D.N. *Geology of India*, London, 1953, p. 385.
55. Krishnan, M.S. *Geology of India and Burma*, Madras, 1960.
56. Glennie, E.A. "Gravity Anomalies in the Structure of Earth Crust", *Memoirs of the Geological Survey of India*, Professional Paper No. 29, Dehradun, 1932, p. 22.

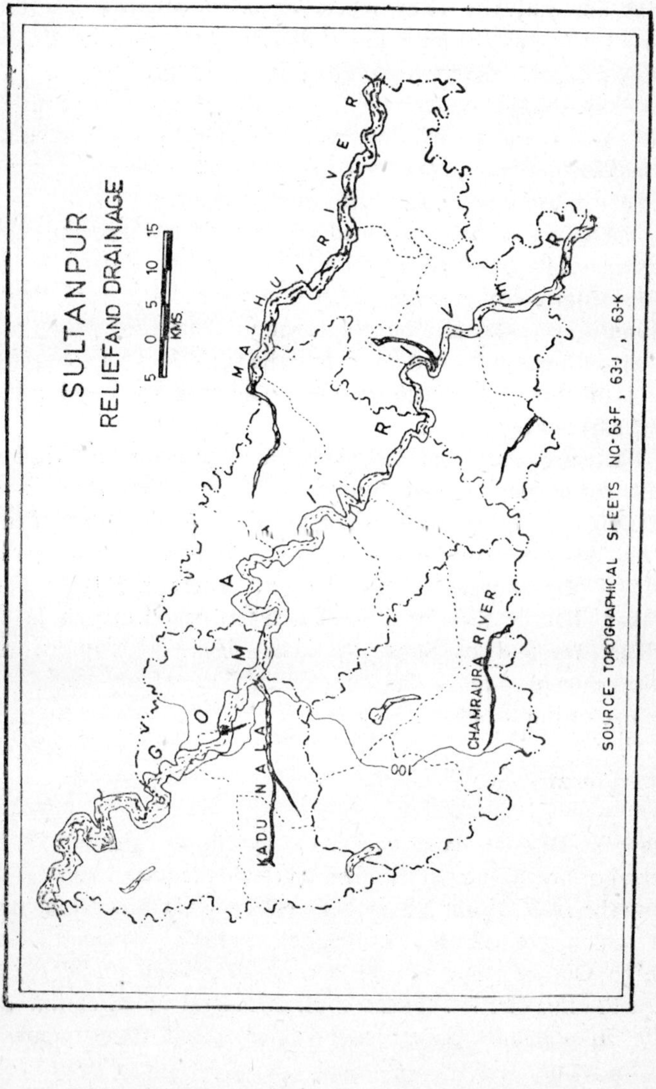
SULTANPUR
RELIEFAND DRAINAGE
5 0 5 10 15
KMS
MAJHUI RIVER
GOMATI RIVER
KADU NALA
CHAMRAURA RIVER
100
SOURCE- TOPOGRAPHICAL SHEETS NO- 63-F, 63-J, 63-K

MAP 3

stream. Wide flood plains and high banks are the common features in the course of the Gomati along with the sand, silt and clay deposits. In general the drainage pattern is dendritic and the general characteristic features available throughout the plains are that the river meets at accute angles and several nalas form parallel or sub-parallel lines to the main stream. The floods are a recurring phenomena in the district. The river has shifted its course frequently covering wide areas throughout the historical times as is reflected by the remnants of their former beds in the form of oxbow lakes, meander loops, dead arms, chaurs, tals and ruins of settlement, etc.

There are widespread arid plains and swampy jhils and marshes. The chief river, Gomati, enters the district at its north-western corner and, after flowing in an exceedingly tortuous, south-easterly course through the centre, passes out at the south-east. Its bed lies below the surface of the country and is at first badly defined but high banks are found in the latter parts of its course.

There are several small streams, the chief being the Majhui which forms part of the boundary between Faizabad and Sultanpur. Another river which flows in Amethi Tahsil of the district, north-west to south-east, is Chamraura. Nala of the Kadu is the longest in the district. There are lakes like Rajakar, Bandh, Katara Ranilake, a few near Kutharu, lakes of Kurahwa and of Sona Mahona. There are a number of shallow jhils or swamps; but none of them are of any considerable size or importance.

1.10.5 Climate

Sultanpur District experiences a rhythm of seasons with reversal of prevailing winds. The reversal is regularly felt twice during the year From November to May winds flow from the west which are of the continental character whereas, from June to October they are of oceanic origin and are felt from the east in this district. This complete reversal of direction of winds on a gigantic scale between summer and winter is called the Monsoon.

1.10.5 (a) Temperature: The monsoon currents are the outcome of the temperature, since it tends to set up surface pressure differences resulting in a low pressure over the land and high pressure over the adjoining areas and vice versa. Hottest months of the year are May and June. The mean temperatures for these months are 32.3° and 31.7°C respectively. The extreme maximum for the month of June is 45.0°C. This is due to the sun's rays falling almost vertically in this month. Conversely the extreme minimum in the month of December is 4.4°C, since the sun is low during this month owing to the earth's position in winter solstice. The annual range of temperature has varied from 16.0°C to 18°C during the past 10 years. Thus the temperature in the district never falls to the freezing point nor is it too high to stop all outdoor activity during the hot weather.

1.10.5 (b) Pressure and Winds: The pressure is 985 millibars during the months of June, July and August. In the months of December and January when the temperature is very low the pressure rises to 1,004 millibars. A shallow cell of high pressure with seaward gradient develops over the land in December and January. In summer the pressure gradient is towards the land. The shallow depressions visiting the district in a cyclonic form ascend from the Bay of Bengal. But the high pressure cells in December and January reach here from West producing rough weather and little rain. These high pressure cells are in cyclonic form and are called the western disturbances.

1.10.5 (c) Moisture and Humidity: The principal factor which determines the potentiality of rainfall is the amount of humidity present in the air. The humidity is high in the months of June, July, August and September. The temperature during these months is high and the pressure is subsequently low. The winds are dry during the months of March, April and May and the relative humidity is comparatively less. Thus in the district, temperature, relative humidity and rainfall are high in the months of June, July, August and September while the remaining months of the year are almost dry.

1.10.5(d) Rainfall: Ninety per cent of the annual rainfall is received only within a short period of four months, i.e. from mid June to mid September. A major part of the year, i.e. from October to June, remains dry since the amount of rainfall is very scanty and sporadic. The annual normal rainfall of the district is 1,000 milimetres but it is very unevenly distributed throughout the year.

(i) Annual Variability of Rainfall: The high percentage of annual variability of rainfall is indicative of the fact that the prosperity of the farmers largely depends on the uniform distribution of annual rainfall, since an annual variability of 12% or more makes an area susceptible to famine.[57] In this respect Sultanpur District stands very often in the danger of super abundance or deficiency of rainfall. There are some places in the district which suffer drought while many others are liable to face serious flood and waterlogging conditions due to heavy rain during the same year. Therefore the annual variability of rainfall is said to affect the cropping system very seriously in the district.

(ii) Seasonal and Monthly Variability of Rainfall: Seasonal and monthly variability of rainfall is rather more significant than the annual variability from the view point of agricultural operations. The amount of annual rainfall may go below or above the average but if it is evenly and timely distributed, the crops are not much affected If the amount of rainfall is insignificant or excessive in the month of June it can delay the sowing of the early kharif crops and thus reduce the returns. If the rainfall is heavy and alternates with long intervals, the vicissitudes of floods and droughts are experienced simultaneously. The state of abnormality frequently causes a great havoc and the whole cropping pattern is seriously affected. Early or late retreat of the summer in September and October affects the maturity of the kharif crops on the one hand and the sowing of the rabi crops on the other. Heavy rainfall in these months may lead to waterlogging while premature sessions of rains may cause postponement or restriction of the sowing of the rabi crops.[58]

Thus taking all the weather elements together and considering the variations in their incidence, the year is divided into three distinct seasons, namely:

57. Blandford, H.F. "Rainfall of India". *Memoirs of the Indian Meteorological Department* Vol. III, 1885-86, p. 130.
58. Clark, K.G.T. "The vicissitudes of the summer rainfall of the Indo-Gangetic Plain and the valley", *Geography*, Vol. XVII, Pt. IV, 1932, p. 288.

(1) The season of rain—mid June to October.
(2) The cold weather season—November to February.
(3) The hot weather season—March to mid June.

Thus we find that the climate of the district is mild and healthy. West winds prevail from October to June, gradually increasing in strength as the hot season approaches. Over the whole district annual rainfall or, in terms of seasons, the northern portion of the district receives more rainfall than the south.

1.10.6 Soils

Sultanpur District consists of soils transported from the sedimentary rocks of the Himalayas and the products of the decomposition of these rocks have been deposited by water action over this tract Therefore the soils of the district fall under the alluvial group of the Indo-Gangetic plain. Scientific data regarding the soils of the district are not available. A brief account of the soils and land classification based on the texture of the soils is however prepared by the revenue authorities to assess the revenue of the land. The main factors governing the system of classification are texture, colour, availability of water and the level of the land.[59] On the basis of the texture the soils of the district have been sub-divided into the following sub-divisions:

(*i*) *Bangar Soil*: The bangar (upland plain) soils vary from sandy loam to stiff clay depending upon topography and drainage. The bangar lands raised up in conformity with the principle that when a river grows older its deposits subsequently occupy a lower position along its basins and the terraces and the flood plain are gradually left free by water spreading action during its old age. These raised uplands which are called bangar are again eroded by the newer channels, but they always remain above the flood plain and are never affected by it. Such land types are observed only in the northern portions of the district. The bangar soil is divided into

59. Shafi, M. *Land Utilization in Eastern Uttar Pradesh*, London, Ph.D. Thesis Published, A.M.U. 1960, pp. 40-41.

various types depending on its texture, consistency and colour.

(*ii*) Loamy Soil: This soil is locally called "Domat Soil", which is extensively distributed over the bangar or upland plain. Patches of loamy soil are noticeable in all tahsils of the district, not in one portion but are spread all over the district. The patches of the Domat soil are very well drained. The content of sand and clay in this soil is almost equal in proportion and therefore can very easily be recognised. Depending on the fertility, the Domat soil has been further divided into (1) Domat Avval, (2) Domat Doam, (3) Domat Soam. According to the percentage of the sand, the colour varies from yellow to brown. The sub-soil is brownish and yellow sometimes mixed together and sometimes alternated with their layers. The water-holding capacity also varies according to the texture. The loamy soil is very much preferred by the cultivators since it is used for double and multiple crops.

(*iii*) Clayey Loam: The clayey loam soil is locally known as "Matiyar soil" and has grey or yellowish grey colour at the surface. The sub-soil at the lower horizons is dark grey in colour. In comparison to the Domat soil the percentage of clay is high, water retentive capacity is also very high. It was recorded at the time of field study that this soil is found almost in every village. It is observed that Matiyar soil occupies low level areas which results in the accumulation of rainwater and therefore the soil responds well to paddy crop. Sometimes this soil is characterised by the presence of calcareous concretions which are leached from the surface and accumulated at various depths in the form of kankar.

(*iv*) Clayey Soil: This soil is locally known as "Dhankhar Soil" and is grey to dark grey in colour. The Dhankhar soil, as its name indicates, is largely used for growing dhan, i.e. paddy. It is a compact and cloddy structure and becomes sticky when wet and very hard when dry. It is mostly used for raising a single crop of late paddy and during other parts of the season it is almost left idle. The villagers call such type of fields with these soils as "Jadhan's Khet" (type of paddy land). In fact ploughing and sowing are almost impossible when it is dry. In Sultanpur District clayey soil is found only in small patches. It mostly lies in the vicinities of tals and jhils and occasionally near low-lying areas.

(*v*) Sandy Soils: It is locally called "Balluo" and "Baldhosari" according to the percentage of sand present in the soil. The water retentive capacity of the soil is very low. Such soils are distributed near river Gomati. The rainwater soaks down very easily due to its coarse texture. The good soil away from the riversides produces a good rabi harvest when very well manured and irrigated. Millet and arhar are sown during the Kharif season.

(*vi*) Khadar Soil: Khadar Soil is mainly confined to the flood plains of the rivers and jhils and is almost annually renewed by the seasonal overflows. The Khadar soil in generally ligh but rich in calcareous matter. The beds of sand and clay are deposited in the alternate layers. The soil in the river vicinities has more sand than that lying away from the river beds. At several places, the soil is loamy and makes cultivation unprofitable. The areas of the Khadar soil are relatively low-lying areas and are subject to occasional flooding and therefore get freshly deposited silt which enhances the fertility of the Khadar soil and permits the growing of excellent rabi crops.

(*vii*) Saline and Alkali Soil: Those soils which have excessive amounts of soluble salts and/or exchangeable sodium, due to which crop production is seriously hindered, are locally known as "Reh/usar". It has high *p*H value of the soil. There is a deficiency of calcium, nitrogen and micro-nutrients like zinc, extremely low water intake rate and hydraulic conductivity resulting in poor moisture transmission characteristics. The presence of calcium carbonate nodules called Kankar, often occurring at about one metre depth, and poor physical condition leads to hindrance in root penetration. Such type of soils are mainly observed in Bhadar, Amethi, Bhnetua Blocks of Amethi Tahsil, Dhanpatganj and Kurebhar, Jaisinghpur Block of Sadar Sultanpur Tahsil and Akhandnagar, Dostpur Blocks of Kadipur Tahsil.

1.10.7 Natural Vegetation

Sultanpur District has an almost inhabited human occupance

for over three millenia, and centuries of plough and pastoral culture has reduced the natural vegetation except in the pockets of the Tarai and some river banks. With a moderate and healthy climate and fertile soil, the district flora presents no peculiarities. The jungles are mainly found in Ramnagar in the south-west though a few patches of Dhak (Butea frond osa) are found elsewhere. Sultanpur is however well wooded and contains magnificent groves of mango, jamun (*Eugenia jambolana*) and Mahua (*Bassia latifolia*).

1.11 FORMAT OF THE STUDY

The present study has been divided into eight chapters, keeping in view the nature of the problem.

In the first chapter the problem of the study has been introduced having a broad literature survey. In this chapter an objective hypothesis. choice of the study area, indicators and time-period, data base, methodology and geographic background of the study are discussed.

In the second chapter classification and definition of all waste land types are discussed, and distribution of all waste land type is analysed for the period 1981-82 and 1977-78 at Tahsil/Block level while for the year 1981 82 Village level distribution has also been discussed. The growth of all waste land types is analysed from 1977-78 to 1981-82 at Tahsil and Block levels.

An attempt has been made in the third chapter to explain the factors influencing all waste land types in which major environmental (morphometric, climatic, hydrologic, pedogenic) and anthropogenic factors are chosen as indicators. In this chapter relationship of all factors with various waste land types and variation in distribution of each waste land type are explained through correlation matrix and stepwise regression analysis respectively.

The genesis of each waste land type has been explained in the fourth chapter.

The fifth chapter is attributed to explain the limitations of each waste land type.

Focus in the sixth chapter is on a discussion of suitable remedies for the reclamation of each type of waste land.

In the seventh chapter economics of each type of waste land reclamation is explained.

The eighth and last chapter of the study provides the major findings of the study on the genesis and utilisation of waste lands of Sultanpur District.

2

TYPOLOGY AND DISTRIBUTION OF WASTE LANDS

2.1 INTRODUCTION

The concept of waste land is not a new phenomenon. The term waste land was used to indicate little used common land usually on less fertile soil, which failed to yield a return to the medieval cultivators. Waste land means differently to the layman. He may define it as land which is lying uninhabited. uncultivated and left after use or land which is no longer serving any purpose. The adjective waste has now disappeared because in many cases these common lands are much valued as open spaces. Dudley Stamp gave an appropriate definition in this connection, "the waste land may be defined as the land which has been previously used but which has been abandoned and for which no further use has been found".[1]

The Waste Land Survey and Reclamation Committee[2] has also defined "the term waste lands as those lands which are either not available for cultivation or are left out of cultivation without being cultivated like fallows and culturable waste". Thus it embraces the following types of waste lands:

(a) Land not available for cultivation, barren and uncultivable waste.

1. Stamp, L,D., *The Land of Britain, Its use and misuse*, 1948, p. 433.
2. Ministry of Food and Agriculture, Government of India, Waste Land Survey and Reclamation Committee, "Report on the Location and Utilization of Waste Lands in India", Part X, Uttar Pradesh, New Delhi, 1961.

(b) Other uncultivated land excluding fallows, culturable waste, permanent pastures and land under miscellaneous trees and crops.

(c) Fallows.

Thus the abovementioned Waste Land Survey and Reclamation Committee includes in the culturable waste all land available for cultivation but not taken up for cultivation or abandoned after a few years owing to some reason or the other. Such lands may be fallow for more than five years and may be covered with shrubs. Lands which are reserved for pasture for the grazing purposes are not included under culturable waste. Fallows of one year are classified as current fallow and those of one to five years are classified as fallows other than current fallow. Thus the waste lands for the purpose of extension of cultivation are confined to other uncultivated land excluding fallows and fallows other than current fallow[3].

Thus the waste lands are such lands as are uncultivable or presently unutilised but have been used previously, which have been abandoned and no further use has been found due to some reason or the other.

2.2 TYPOLOGY OF WASTE LANDS

The agricultural census land use classification considers mainly five categories of land use which are: forest, land not available for cultivation, other uncultivated land excluding fallows, fallows and net cultivated land.

From these five types of land use three categories, viz. land not available for cultivation, other uncultivated land excluding fallows and fallows are helpful in formulating waste land categories. These categories are sub-divided again and from this, only culturable waste and fallows other than current fallow are taken for further classification of waste land types. There are two categories in the land not available for cultivation, firstly, land put to non-agricultural use which cannot be categorised as waste

3. Shafi. M., "The Problems of Waste Lands in India", *The Geographer*, Vol. XV, November, 1968, A.M.U. Geographical Society, Aligarh.

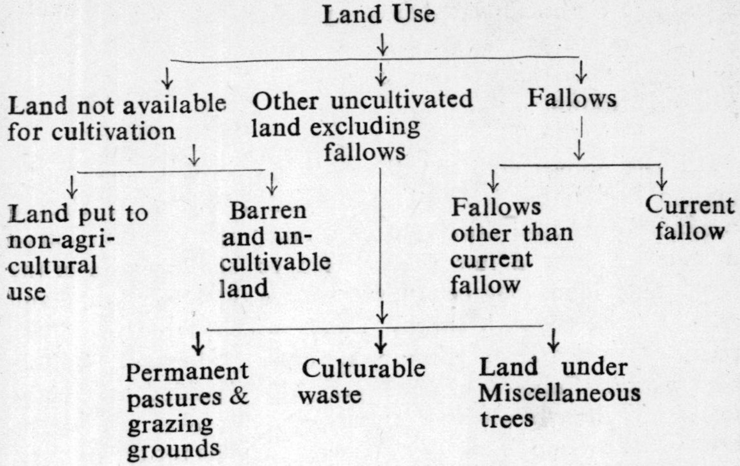

land because such type of lands are used not for agricultural use, secondly, the barren and uncultivable land cannot be brought under cultivation because lands are left permanently out of cultivation. The other uncultivated lands, excluding fallows, permanent pastures and grazing grounds, land under miscelleneous trees, are used for some or the other purpose but not for any agricultural purpose, so these cannot be categorised under waste land while from this category cultivable waste land has been taken for further classification of waste land because such type of lands are available for cultivation as these lands were previously under cultivation but presently are not due to some reason or the other. The current fallow or the fallow land is not considered for further classification of waste lands because this is a land which is left fallow to maintain the fertility status of the soil, and within a year the same fields are recultivated but fallows other than current fallow are taken for further classification of waste lands because after one year if the same lands are not brought back under cultivation due to some problems of the farmers, the same may turn into waste lands.

The Milan Khasara of the village or the Land Revenue Records of Sultanpur District classifies waste lands into following types: (a) Usar Land, (b) Banjar Land (more than 5 year),

(c) Old Fallow (2-5 years), (d) Fallow (1-2 years), (e) Other types of waste land.

However there are certain other types of waste lands which are evidenced in the villages of Sultanpur District and it will be useful to include them in a classification of waste lands. Thus the revised classification of waste land takes the following form:

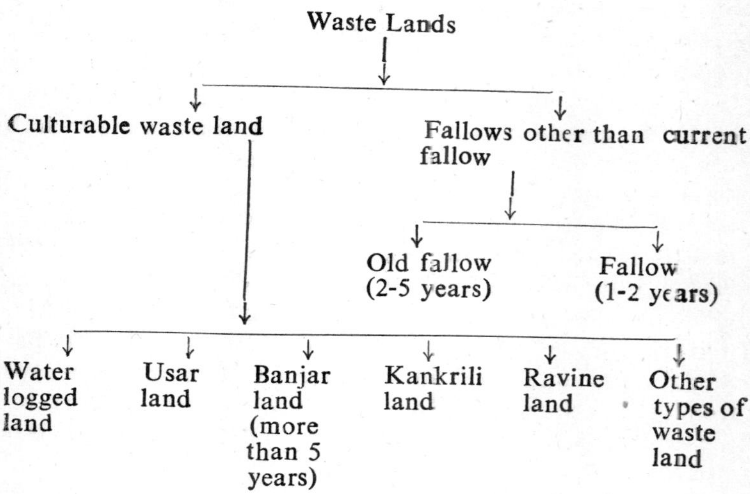

These different categories of waste lands are explained below:

2.2.1 Waterlogged Land

Waterlogged lands are those areas where wet lands are unable to bear the weight of tractors and cannot be stocked for fear of lasting damage to the soil texture. These are characterized by high water table. It can also be said that waterlogging usually occurs due to the rise of sub-soil water table and this happens in almost all the overirrigated areas. Plants derive the bulk of their water requirements through their root system and so water must be available in the soil in enough quantities needed by the plant. Too little water in the soil will result in the withering of the plant and too much water; on the other hand, will cause waterlogging.[4]

"Land is said to be waterlogged when the soil pores within the root zones of the crop get saturated with water, cutting the normal circulation of the air which is very necessary for the growth of the plants and reduces yield of the crops.[5]

Singh[6] has said that in the pre-canal period the water was in a state of equilibrium. The inflow into the soil from the outer sources was equal to the outflow due to seepage. It should not be concluded from this that the sub-soil water table remained static. It did fluctuate with the amount of rainfall from year to year but there was no continuous rising trend. There could be a sudden rise in the year of heavy rainfall which would be adjusted in a subsequent dry year.

In the study area the rising trend in the water table is recorded with the introduction of tube-well irrigation. Water table is found very low near the tubewell areas but in other areas it is recorded high due to seepage and outflow of water. Subsequently after the tube-well irrigation, when canal irrigation is introduced a rising trend in the sub-soil water table has been recorded due to the addition of the water through seepage from canal. Near the Gomati sub-soil water table is recorded high due to the encroachment on the flood zones. It is also observed that water seepage did occur through the unlined canal banks. Added to this there is also a misconception among the farmers that higher doses of irrigation water would be helpful in getting higher yields. This results in a heavier irrigation by which more water seeps through and adds to the sub-soil water table. In most of these areas it has been observed that there is no outlet which increases correspondingly with the rising water table resulting in waterlogging conditions.

Thus waterlogging is the cause of the rising trend in the sub-soil water table due to either over-irrigation or accumulation of rainywater in the low-lying areas. This may also be due

4. Symon, L., *Agricultural Geography*, 1968, London, p. 21.
5. Committee on Natural Resources, Planning Commission (New Delhi), *Study on Waste Lands, including Saline, Alkali and Waterlogged land and their Reclamation measures*, 1963, pp. 105-127.
6. Singh, A.L., *Economics and Geography of Agricultural Land Reclamation*, New Delhi, 1978, pp. 14-15.

to the seepage from the unlined canal banks and more outflow from the irrigated fields, and such water inhibits the growth and activity of plants and due to which waterlogged lands are left uncultivated.

2.2.2 Usar Land

Those fluffy soils which have a whitish or greyish appearance and are deposited on the surface are generally known as *reh* or *rehar* or *usar* in Uttar Pradesh.

Govinda Rajan[7] and others have defined usar lands as "those lands which have an excessive amount of soluble salts and/or exchangeable sodium due to which crop production is seriously hindered are often referred to as salt affected soils and by local name are known as usar lands."

The saline and alkali soils occur most commonly under arid climates. They owe their distinctive character to the fact that they contain excessive concentration of soluble salts or exchangeable sodium or both.[8]

The low permeability of the soil causes poor drainage by impeding downward movement of water. Low permeability may be the result of an unfavourable soil texture or structure or presence of indurated layers. The latter may consist of a clay pan and as a result of excessive irrigation, the water table rises and the non saline soil becomes saline.[9]

The injurious salts in the saline and alkali soils, which are the decomposition products of igneous rocks under natural weathering processes, are present in the soil profile itself and they become evident in the surface layers under certain predisposing conditions like: (i) Arid or semi-arid climate; (ii) Impervious sub-soil or hard pan; and (iii) Temporary abundance of humidity in the soil interspersed with dry periods.[10]

7. Govinda Rajan, S.V. and Gopala Rao, H.G., *Studies of Soils of India*, New Delhi, 1978, p. 373.
8. Dhawan, C.L., *Land Reclamation*, Central Board of Irrigation and Power, Publication No. 43, Vol. III, p. 1.
9. Signond, A.A. Joe, "The Alkali Soils in Hungary and Their Reclamation Measures", *Soil Science*, 18, 1924, pp. 379-381.
10. Leather, J.W., "Soil", in the Agricultural Ledger, 3, *Agricultural Soils*, 1898, p. 24.

Usar land is the plantless land which has whitish-greyish or blackish layers on the sub-soil and at the surface. This is not easily cultivable. Sometimes it is called as reh or kallar because it has an excessive amount of soluble salts.

2.2.3 Banjar Land

Banjar lands are these lands which were taken up for cultivation but are temporarily left out of cultivation for a period of not less than five years. These lands are left uncultivated for a very long period of time and such type of lands are called by villagers as Banjar land.

2.2.4 Old Fallow Land

Old fallow lands are those lands which were under cultivation but presently are out of cultivation for a period of not less than two years and not more than five years. The reasons for leaving such cultivable land out of cultivation as recorded from field observations are the poor fertility status of the soil, other natural calamities and socio-economic factors.

2.2.5 Fallow Land

These lands which would be taken up for cultivation but are temporarily out of cultivation for not less than one year and not more than two years are called fallow land. The reasons for leaving such cultivable land as fallow are low fertility status of the soil, socio-economic conditions of the farmers which do not permit farmers to bring back such lands under cultivation, inadequate water supply and low level of agricultural technology besides other environmental factors which also add to the reasons mentioned above for these lands to remain fallow.

The duration of time is the main criterion in the classification of fallow, old fallow and banjar lands, as the cultivating conditions deteriorate in proportion to the length of time for which the land is left uncultivated. Mainly the fertility status of the soil gets more deteriorated in the case of banjar land and less in case of fallow land, and as it will be seen in the discussion of distribution pattern of these waste lands, there is no marked regular pattern shown by these lands.

2.2.6 Kankrili Land

The land where concretion takes place and kankars are visible on the upper earth surface and sometimes at a certain depth is called kankrili. In many areas the kankars are exposed on the surface mainly because of the removal of the top soil due to erosional activities.

2.2.7 Ravine Land

A ravine is a deep narrow gorge, worn out by a stream or torrent of water. According to Moore[11] "Ravine is a long narrow depression in the earth's surface, rather smaller than a valley but larger than a gully. Several gullies often lead to a ravine and several ravines to a valley."

Sharma[12] has defined "Ravine as a channel of flow denuded and guided essentially by the process of rejuvenated streams having steep sides and head scarps with a width always greater than a gully".

The world ravine denotes gullied land containing a system of gullies running more or less parallel to each other and entering a nearby river flowing much lower than the surrounding table lands.[13]

Thus a ravine means a system of gullies or gorges worn out by torrents of water and gullies running more or less parallel to each other and draining into a major river or its tributaries.

2.2.8 Other Types of Waste Land

Other types of waste land include those lands which cannot, as predefinition, be categorised into the above heads. This is primarily a class derived from the village handbook (Milan

11. Moore, W.G., *A Dictionary of Geography*, 1976, Penguin Books, p. 182.
12. Sharma, H.S., *Ravine Erosion in India*, 1980. Concept Publishing Co., New Delhi.
13. India, Planning Commission; Natural Resource Committee, 1965, "Study on Survey and Reclamation of Ravines in India", Planning Commission, New Delhi.

Khasara). In this category can be classified such lands as are primarily out of cultivation at present.

2.3 Distribution of Waste Lands

It was indicated earlier that the area under waste lands in Sultanpur District is quite big (12.85% of the total area of the district in the year 1977-78). The distribution of waste lands in 1981-82 seems to have increased by 10.93% from that in 1977-78, the present (1981-82) distribution percentage being 13.72%. The distribution and growth of different waste land types are as follows :

TABLE 2.0

Waste Lands in Sultanpur District

Waste Land Types	% to total area of the district in 1977-78	% to total area of the district in 1981-82	% growth from 1977-78 to 1981-82
1. Waterlogged land	0.43	0.58	29.96
2. Usar land	0.46	0.54	15.79
3. Banjar land	3.93	4.18	10.48
4. Old Fallow land	1.35	1.51	11.68
5. Fallow land	2.73	2.94	7.92
6. Kankrili land	0.06	0.07	15.50
7. Ravine land	0.43	0.53	16.47
8. Other types of waste land	3.14	3.43	9.03
Total Waste lands	12.85	13.72	10.93

The distribution of waste lands according to the categories (Table 2.0) indicates that the banjar land formed substantially a high proportion (3.93%) in 1977-78, followed by other types of waste land (3.14%), fallow land (2.73%), old fallow land (1.35%), usar land (0.46%) and ravine and waterlogged land (0.43%) whereas the kankrili land is only nominal being less than 0.1%.

It can also be seen in the table that the banjar land formed a high proportion (4.18%) of the total area of the district in the year 1981-82 followed by other types of waste land (3.43%), fallow land (2.94%), old fallow land (1.51%), waterlogged land (0.58%), usar land (0.54%) and ravine land (0.53%), whereas the kankrili land is only nominal being less than 0.1%.

A comparison would show that the waste lands have increased almost in all categories during the 1977-78 to 1981-82 period, the increase being the highest in the waterlogged land (29.96%) followed by ravine land (16.47%), usar land (15.79%), kankrili land (15.50%), old fallow land (11.68%), banjar land (10.48%) and other types of waste land (9.03%), lowest growth being that of fallow land (7.92%). Thus there is a significant change in all types of the waste land in Sultanpur District.

The distribution of waste lands in 1977-78 (Table 2.1) at tahsil level is observed to be high (14.97%) in Sadar Sultanpur Tahsil and low (9.60%) in Musafirkhana Tahsil. The waterlogged land is found to be high (1.26%) in Sadar Sultanpur and low (0.03%) in Amethi Tahsil. The usar land is found to be high (0.67%) in Kadipur and low (0.14%) in Musafirkhana Tahsil. The banjar land is recorded to be high (4.74%) in Sadar Sultanpur and low (2.75%) in Amethi Tahsil. Sadar Sultanpur also ranks high (1.70%) in old fallow land distribution and the lowest (0.95%) range is found in Amethi Tahsil. Fallow land is found more (3.44%) in Amethi Tahsil and less (1.48%) in Musafirkhana Tahsil. The kankrili land is distributed as (0.22%) in Sadar Sultanpur and 0.02% in Kadipur Tahsil. The ravine land is recorded high (0.66%) in Sadar Sultanpur and low (0.19%) in Kadipur Tahsil. The other types of waste land are found to be high (3.48%) in Kadipur and low (1.44%) in Musafirkhana Tahsil of the district.

In the year 1981-82 (Table 2.2) the distribution pattern of total waste land is same as in 1977-78 which is found to be high (16.20%) in Sadar Sultanpur and low (10.95%) in Musafirkhana Tahsil. Distribution of waterlogged land is high (0.78%) in Kadipur and low (0.04%) in Amethi Tahsil. The usar land is recorded to be more (0.75%) in Sadar Sultanpur and low (0.20%) in Musafirkhana Tahsil. The banjar land also shows a similar distribution pattern as that of 1977-78 (5.16%) in Sadar

TABLE 2.1 : **Waste Lands in 1977-78**

(area in acres)

Blocks/Tahsils	Total Area	Water-logged Land	Usar Land	Banjar Land	Old Fallow Land	Fallow Land	Kankrili Land	Ravine Land	Other Types of Waste Land	Total Waste Land
1	2	3	4	5	6	7	8	9	10	11
Jagdishpur	73211	316 (0.43)	167 (0.23)	3416 (4.67)	845 (1.15)	267 (0.36)		693 (0.95)	1374 (1.88)	7078 (9.67)
Gaura Jamo	61763	1017 (1.65)	134 (0.22)	1781 (2.88)	719 (1.16)	1637 (2.65)			1185 (1.94)	6473 (10.48)
Musafirkhana	55782	86 (0.15)	76 (0.13)	1923 (3.45)	681 (1.22)	723 (1.30)		765 (1.37)	354 (0.63)	4578 (8.21)
Baldirai	56238			2531 (4.50)	456 (0.81)	1017 (1.81)		934 (1.66)	638 (1.13)	5576 (9.92)
Tahsil Musafirkhana	246974	1419 (0.57)	377 (0.14)	9651 (3.91)	2701 (1.09)	3644 (1.48)		2392 (0.97)	3551 (1.44)	23735 (9.60)
Amethi	40520	68 (0.17)	138 (0.34)	1012 (2.50)	253 (0.62)	1217 (3.00)			831 (2.05)	3459 (8.64)
Bhadar	52061		732 (1.41)	1968 (3.78)	638 (1.23)	1964 (3.77)			1538 (2.95)	6840 (13.14)

1	2	3	4	5	6	7	8	9	10	11
Bhnetua	61974		264 (0.43)	2053 (3.31)	651 (1.05)	1836 (2.96)			2431 (3.92)	7235 (11.67)
Gauriganj	67505	13 (0.02)	28 (0.04)	1374 (1.89)	523 (0.77)	2613 (3.87)			2124 (3.15)	6375 (9.44)
Tahsil Amethi	222060	81 (0.03)	1162 (0.52)	6107 (2.75)	2065 (0.95)	7630 (3.44)			6924 (3.12)	23969 (10.81)
Kadipur	64685	213 (0.87)	22 (0.03)	2164 (3.35)	1835 (2.84)	1912 (2.96)	29 (0.03)	84 (0.13)	2451 (3.79)	8710 (13.39)
Lambhua	40037	207 (0.52)	451 (1.13)	1538 (3.84)	394 (0.98)	2136 (5.34)		49 (0.24)	1754 (4.38)	6780 (16.93)
Pratappur Kamaicha	47315	132 (0.28)	55 (0.12)	1453 (3.07)	933 (1.97)	584 (1.23)	7 (0.01)	21 (0.04)	1023 (2.16)	4207 (8.89)
Dostpur	67777	544 (0.87)	845 (1.25)	2450 (3.61)	1214 (1.79)	1416 (2.09)	38 (0.06)	138 (0.20)	3324 (4.90)	9879 (14.68)
Akhandnagar	64152	356 (0.55)	52 (0.08)	2036 (3.17)	458 (0.71)	1085 (1.69)			5136 (8.01)	9123 (14.22)
Tahsil Kadipur	283966	1922 (0.47)	1425 (0.50)	9641 (3.40)	4834 (1.70)	7133 (2.51)	78 (0.02)	243 (0.19)	13688 (3.48)	38556 (10.72)
Jaisinghpur	52206	158 (0.30)	381 (0.73)	2431 (4.66)	578 (1.11)	1076 (2.06)	90 (0.19)	308 (0.59)	1673 (3.20)	6705 (12.84)

TABLE 2.1 (*Contd.*)

Dhanpatganj	55425	136 (0.25)	305 (0.55)	2614 (4.72)	817 (1.47)	1214 (2.19)	141 (0.25)	382 (0.69)	1265 (2.65)	6874 (12.40)
Kurebhar	54384	415 (0.76)	140 (0.28)	1018 (1.87)	1382 (2.54)	1938 (3.56)	121 (0.22)	285 (0.52)	1823 (3.35)	7168 (13.18)
Kurwar	53749	25 (0.05)	58 (0.11)	2521 (4.69)	213 (0.40)	568 (1.06)	143 (0.27)	13 (0.77)	542 (1.01)	4483 (8.34)
Dubepur	45469	105 (0.23)	392 (0.86)	1892 (4.16)	456 (1.00)	1545 (3.40)	38 (0.08)	83 (0.18)	780 (1.72)	5291 (11.64)
Bhadainya	52105	654 (1.26)	452 (0.87)	1914 (3.67)	689 (2.32)	2914 (5.59)	43 (0.08)	263 (0.58)	1652 (3.17)	8581 (16.47)
Tahsil Sadar Sultanpur	261233	1493 (0.57)	1738 (0.67)	12690 (4.74)	4135 (1.58)	9255 (3.54)	586 (0.22)	1734 (0.66)	7735 (2.96)	39102 (14.97)
District Sultanpur	1014233	4515 (0.43)	4702 (0.46)	38399 (3.93)	13735 (1.35)	27662 (2.73)	658 (0.06)	4369 (0.43)	31898 (3.14)	125362 (12.85)

Note: The figures in parenthesis represent proportion to the total area.

TABLE 2.2 **Waste Lands in 1981-82**

(area in acres)

Blocks/Tahsils	Total Area	Water-logged Land	Usar Land	Banjar Land	Old Fallow Land	Fallow Land	Kankrili Land	Ravine Land	Other types of Waste Land	Total Waste Land
1	2	3	4	5	6	7	8	9	10	11
Jagdishpur	73211	435 (0.59)	213 (0.29)	3903 (5.33)	1006 (1.37)	372 (0.51)		757 (1.03)	1447 (1.98)	8133 (11.11)
Gaura Jamo	61763	1219 (1.97)	192 (0.31)	1983 (3.21)	805 (1.30)	1838 (2.98)			1298 (2.10)	7335 (11.89)
Musafirkhana	55762	97 (0.17)	90 (0.16)	2025 (3.63)	746 (1.34)	799 (1.43)		879 (1.58)	461 (0.83)	5097 (9.14)
Baldirai	56238			2966 (5.27)	532 (0.95)	1126 (2.00)		1027 (1.83)	829 (1.47)	6480 (11.52)
Tahsil Musafirkhana	246974	1751 (0.71)	495 (0.20)	10877 (4.40)	3089 (1.25)	4136 (1.67)		2663 (1.08)	4035 (1.63)	27039 (10.95)
Amethi	40520	91 (0.22)	141 (0.35)	1042 (2.57)	287 (0.71)	1316 (3.25)			928 (2.29)	3802 (9.32)
Bhadar	52061		876 (1.68)	2791 (5.36)	665 (1.28)	2024 (3.89)			1653 (3.18)	8009 (15.38)
Bhnetua	61974		288 (0.46)	2156 (3.46)	781 (1.26)	2048 (3.30)			2528 (4.08)	7801 (12.59)

TABLE 2.2 (*Contd.*)

Gauriganj	67505	9 (0.01)	36 (0.05)	1774 (2.63)	610 (0.90)	2804 (4.15)			2373 (3.52)	7606 (11.25)
Tahsil Amethi	222060	100 (0.04)	1341 (0.60)	7763 (3.50)	2343 (1.06)	8192 (3.69)			7412 (3.37)	27218 (12.25)
Kadipur	64685	352 (0.54)	33 (0.05)	2318 (3.58)	1926 (2.98)	2018 (3.18)	36 (0.06)	106 (0.16)	2638 (4.08)	9427 (14.57)
Lambhua	40037	276 (0.69)	552 (1.38)	1685 (4.21)	439 (1.10)	2264 (5.65)		136 (0.30)	1866 (4.66)	7082 (17.69)
Pratappur Kamaicha	47315	143 (0.30)	60 (0.13)	1467 (3.10)	1005 (2.12)	699 (1.48)	10 (0.02)	29 (0.06)	1072 (2.27)	4485 (9.48)
Dostpur	67777	949 (1.40)	938 (1.38)	2672 (3.94)	1342 (1.98)	1513 (2.23)	54 (0.08)	177 (0.26)	3475 (5.13)	11120 (16.41)
Akhandnagar	64152	491 (0.77)	63 (0.10)	2153 (3.36)	540 (0.84)	1193 (1.86)			5940 (9.26)	10380 (16.18)
Tahsil Kadipur	283966	2211 (0.78)	1646 (0.58)	10295 (3.63)	5252 (1.85)	7687 (2.71)	100 (0.04)	448 (0.26)	14991 (5.28)	42500 (14.98)
Jaisinghpur	52206	231 (0.44)	414 (0.79)	2533 (4.85)	614 (1.18)	1127 (2.16)	98 (0.18)	328 (0.63)	1763 (3.38)	7102 (13.60)
Dhanpatganj	55425	156 (0.28)	368 (0.66)	2705 (4.88)	906 (1.63)	1322 (2.39)	150 (0.27)	409 (0.74)	1324 (2.39)	7340 (13.24)
Kurebhar	54384	501 (0.92)	150 (0.27)	1274 (2.34)	1483 (2.73)	2086 (3.84)	137 (0.25)	394 (0.72)	1909 (3.51)	7477 (13.75)

Kurwar	53749	29	74	2913	257	669	179	556	651	5328
		(0.05)	(0.14)	(5.42)	(0.48)	(1.24)	(0.33)	(1.03)	(1.21)	(9.91)
Dubepur	45469	134	407	1992	550	1635	47	95	881	5741
		(0.29)	(0.90)	(4.38)	(1.21)	(3.60)	(0.10)	(0.21)	(1.94)	(12.63)
Bhadainya	52105	758	550	2064	846	3000	55	332	1732	9337
		(1.45)	(1.06)	(3.96)	(1.62)	(5.76)	(0.11)	(0.64)	(3.32)	(17.50)
Tahsil Sadar Sultanpur	261233	1309	1959	13481	4656	9839	660	2114	8260	42325
		(0.69)	(0.75)	(5.16)	(1.70)	(3.77)	(0.25)	(0.81)	(3.16)	(16.20)
District Sultanpur	1014233	5371	5441	42416	15340	29854	760	5185	34777	139075
		(0.58)	(0.54)	(4.18)	(1.51)	(2.94)	(0.07)	(0.53)	(3.43)	(13.72)

Note: The figures in parentheses represent proportion to the total area.

Sultanpur and low (3.50%) in Amethi Tahsil. The old fallow land level is seen to be high (1.85%) in Kadipur and low (1.06%) in Amethi Tahsil. The distribution of fallow land is found to be high (3.77%) in Sadar Sultanpur Tahsil and low (1.67%) in Musafirkhana Tahsil. Kankrili land has a high (0.25%) percentage in Sadar Sultanpur and low (0.04%) in Kadipur Tahsil. The ravine land distribution is observed to be high (1.08%) in Musafirkhana and low (0.26%) in Kadipur Tahsil. The distribution of other types of waste land is found to be high (5.28%) in Kadipur and low (1.63%) in Musafirkhana Tahsil.

Observing the growth of waste lands, Sadar Sultanpur records a low growth rate in usar, banjar, fallow and kankrili land. (Table 2.3). Amethi Tahsil has high growth of banjar land and low growth of usar land and other types of waste land. It has been observed that Kadipur Tahsil has the high percentage growth of waterlogged land, kankrili land, ravine land and low growth in old fallow land, and Musafirkhana Tahsil has high percentage growth of waste lands, usar land, old fallow land, fallow land, other types of waste land, and low percentage growth is recorded in ravine land.

The data for the years 1977-78 and 1981-82 is collected for the tahsil and block levels but for the village level it was not possible to do so for the period 1977-78, hence the village level distribution of waste land pertains to only present conditions (1981-82), i.e. the period of the data collection. Since there are wide variations in the distribution of waste lands within the district and at tahsil level the same variations have also been depicted at block level for the year 1977-78 (Table 2.1) and 1981-82 (Table 2.2) and the change over time is depicted in Table 2.3. The variation in the distribution of waste lands and also the change over time is discussed below categorywise for the block level, and the village level distribution pattern of waste lands is discussed for the year 1981-82.

2.3.1. Waterlogged Land

The distribution of waterlogged land in 1977-78 (Map 4) ranges from 0.02% in Gauriganj Block to 1.65% in Gaura Jamo Block as given in Table 2.1. It is seen from Map 4 that there is a

TABLE 2.3 : Percentage Growth of Waste Lands (1977-78 to 1981-82)

(area in acres)

Blocks/Tahsils	Water-logged Land	Usar Land	Banjar Land	Old Fallow Land	Fallow Land	Kank-rili Land	Ravine Land	Other types of Waste Land	Total Waste Land
1	2	3	4	5	6	7	8	9	10
Jagdishpur	37.65	27.54	14.25	19.05	39.32		9.23	5.31	14.90
Gaura Jamo	19.86	43.28	11.34	11.96	12.27			9.54	13.32
Musafirkhana	12.79	18.42	5.30	9.54	9.60		14.90	26.30	11.35
Baldirai			17.18	16.67	10.71		9.95	29.93	11.33
Tahsil Musafirkhana	23.39	30.65	12.70	14.36	13.50		11.32	13.62	14.06
Amethi	33.82	2.17	2.96	13.43	8.13			11.67	9.92
Bhadar		19.67	41.82	4.23	3.05			7.47	17.09
Bhnetua		9.09	5.01	19.96	11.55			3.99	7.71
Gauriganj	—30.76	28.57	21.83	16.63	7.31			11.73	19.16
Tahsil Amethi	2.35	15.40	27.12	13.46	7.36			7.05	13.55
Kadipur	65.25	5.00	7.12	4.96	5.54	24.14	26.19	7.63	8.23
Lambhua	33.33	22.39	9.56	11.42	5.80		24.28	6.39	4.45
Pratappur Kamaicha	8.33	9.09	0.96	7.71	19.69	42.85	38.09	4.79	6.60

TABLE 2.3 (*Contd.*)

1	2	3	4	5	6	7	8	9	10
Dostpur	71.23	11.00	9.06	10.54	6.85	42.10	28.26	4.54	12.56
Akhandnagar	37.92	21.15	5.75	17.90	9.95			15.65	13.77
Tahsil Kadipuri	45.26	15.15	6.78	8.64	7.77	42.85	38.39	9.51	11.13
Jaisinghpur	46.20	8.66	4.19	6.23	4.74	1.11	6.49	5.37	5.92
Dhanpatganj	14.70	20.65	3.48	10.89	8.89	6.38	7.06	4.66	6.77
Kurebhar	20.72	7.12	25.14	7.31	7.64	13.22	38.24	4.71	4.31
Kurwar	16.00	27.58	15.54	20.65	17.78	25.17	34.62	20.11	18.84
Dubepur	27.61	3.82	5.28	20.61	5.82	23.68	14.45	12.94	8.51
Bhadainya	15.90	21.68	7.83	22.78	2.95	27.91	26.24	4.84	8.81
Tahsil Sadar Sultanpur	21.65	12.71	6.23	12.60	6.31	12.62	21.91	6.79	8.26
District Sultanpur	29.96	15.79	10.48	11.68	7.92	15.50	16.47	9.03	10.93

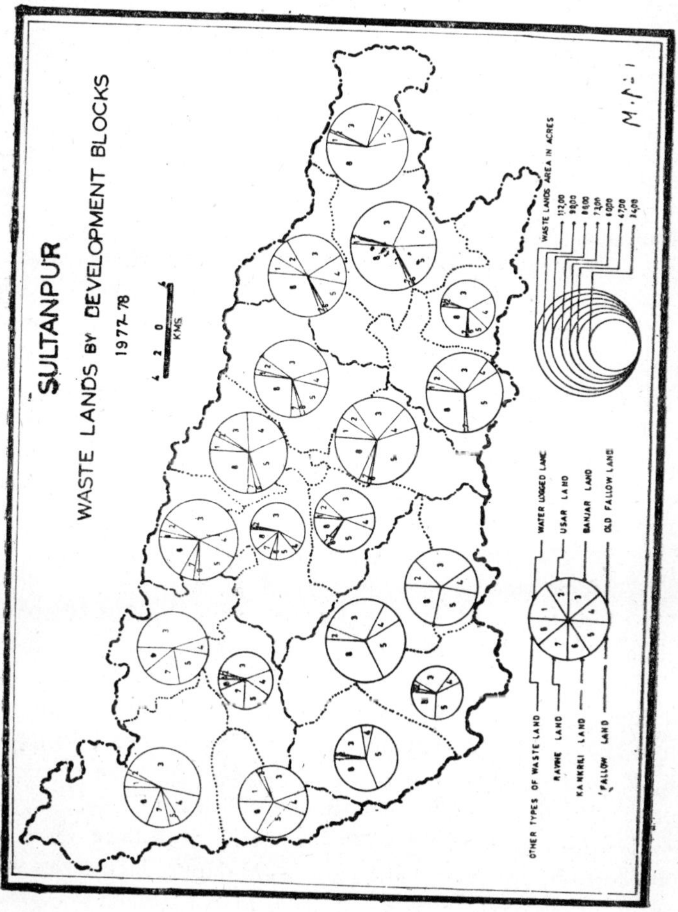
SULTANPUR
WASTE LANDS BY DEVELOPMENT BLOCKS
1977-78
KMS
WASTE LANDS AREA IN ACRES
OTHER TYPES OF WASTE LAND
RAVINE LAND
KANKRILI LAND
FALLOW LAND
WATER LOGGED LAND
USAR LAND
BANJAR LAND
OLD FALLOW LAND

MAP 4

high (more than 1%) percentage of waterlogged land in Gaura Jamo and Bhadainya Blocks and medium (0.50% to 1.00%) percentage of waterlogged land is recorded in Kadipur, Lambhua, Dostpur, Akhandnagar and Kurebhar Blocks and low (less than 0.50%) percentage distribution of waterlogged land is observed in Jagdishpur, Musafirkhana, Amethi, Gauriganj, Pratappur Kamaicha, Jaisinghpur, Dhanpatganj, Kurwar and Dubepur Blocks.

In the year 1981-82 (Map 5) the waterlogged land varies from 0.01% in Gauriganj Block to 1.97% in Gaura Jamo Block which is given in Table 2.2. High percentage of waterlogged land (more than 1%) is observed (Map 5) in Gaura Jamo, Dostpur and Bhadainya Blocks, medium percentage (0.50% to 1%) is seen in Jagdishpur, Kadipur, Lambhua, Akhandnagar and Kurebhar Blocks and low percentage (less than 0.50%) of waterlogged land is found in Musafirkhana, Amethi, Gauriganj, Pratappur, Kamaicha, Jaisinghpur, Dhanpatganj, Kurwar and Dubepur Blocks.

The percentage growth of the waterlogged land (Map 6) is recorded to be the maximum (71.23%) in Dostpur Block and a negative change, from 1977-78 to 1981-82 (—30.76%) of waterlogged land is evidenced in Gauriganj Block where the problem of waterlogging has improved as given in Table 2.3. It is observed from Map 6 that there is a high percentage (more than 50%) of waterlogged land in Kadipur and Dostpur Blocks, medium (25% to 50%) percentage change is seen in Jagdishpur, Amethi, Lambhua, Akhandnagar, Jaisinghpur, Dubepur Blocks and low percentage (less than 25%) growth is observed in Gaura Jamo, Musafirkhana, Pratappur, Kamaicha, Dhanpatganj, Kurebhar, Kurwar and Bhadainya Blocks while negative change is recorded in Gauriganj Block of Amethi Tahsil.

The distribution of waterlogged land at village level is recorded to be very high (21.94%) in Malapur Jagdishpur of Kadipur Tahsil. It is recorded from Map 7 that most of the villages of the district lie in very low (0 to 10%) percentage category of the waterlogged land. Of these 45 villages are in Gaura Jamo of Musafirkhana Tahsil, 35 villages in Dostpur Block of Kadipur Tahsil and 80 villages in Sadar Sultanpur, 64 villages in Musafirkhana Tahsil, 91 villages in Kadipur and very few (nine villages) are in Amethi Tahsil. Low percentage

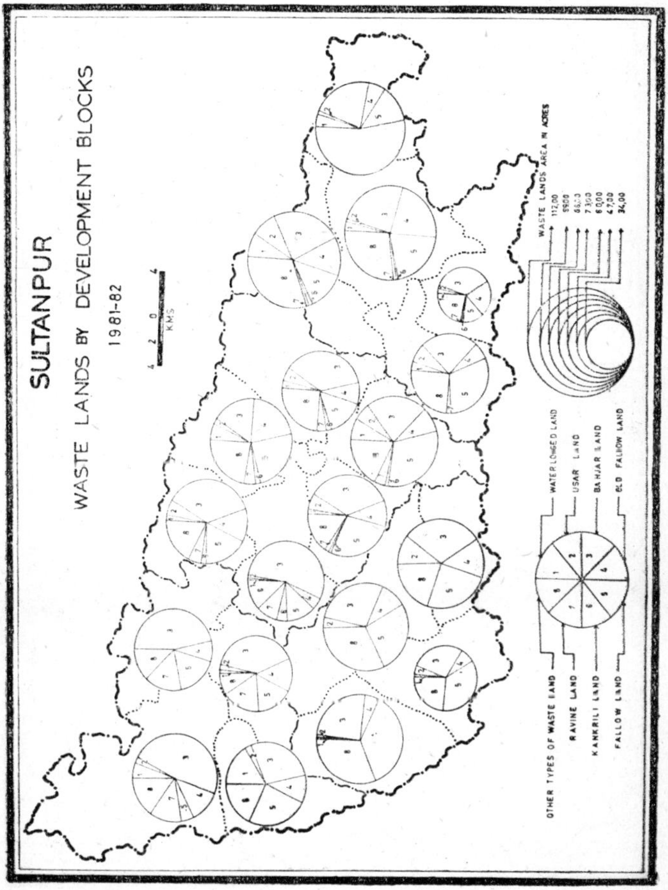
SULTANPUR
WASTE LANDS BY DEVELOPMENT BLOCKS
1981-82
KMS
OTHER TYPES OF WASTE LAND
RAVINE LAND
KANKRILI LAND
FALLOW LAND
WATER LOGGED LAND
USAR LAND
BAHJAR LAND
OLD FALLOW LAND
WASTE LANDS AREA IN ACRES

MAP 5

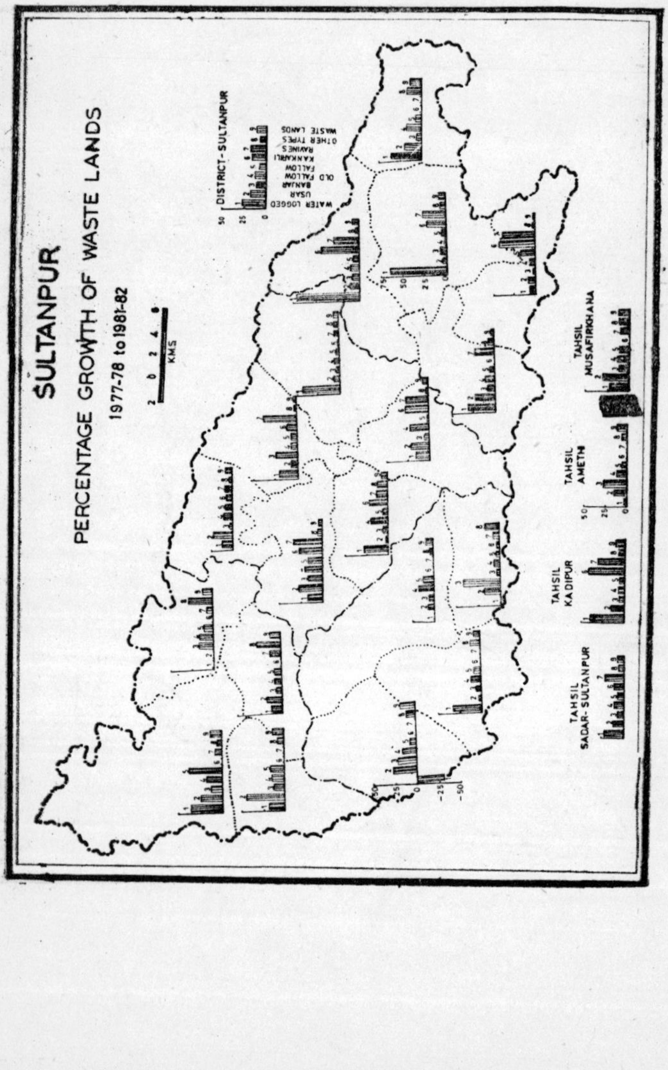
SULTANPUR
PERCENTAGE GROWTH OF WASTE LANDS
1977-78 to 1981-82
KMS
DISTRICT-SULTANPUR
WATER LOGGED
USAR
BANJAR
OLD FALLOW
FALLOW
KANKARLI
RAVINES
OTHER TYPES
WASTE LANDS
TAHSIL SADAR-SULTANPUR
TAHSIL KADIPUR
TAHSIL AMETHI
TAHSIL MUSAFIRKHANA

MAP 6

as given in the Table 2.1 ranges from 1.87% in Kurebhar Block to 4.69% in Jaisinghpur Block of Sadar Sultanpur Tahsil. High percentage of banjar land (more than 3%) is observed in Jagdishpur, Musafirkhana, Baldirai, Bhadar, Bhnetua, Kadipur, Lambhua, Pratappur Kamaicha, Dostpur, Akhandnagar, Jaisinghpur, Dhanpatganj, Kurwar, Dubepur and Bhadainya Blocks as depicted in Map 4. Low percentage (less than 3%) of banjar land is observed in Amethi, Gaura Jamo and Kurebhar Blocks.

Banjar land distribution pattern in 1981-82 varies from 2.34% in Kurebhar Block to 5.42% in Kurwar Block of Sadar Sultanpur Tahsil which is given in the Table 2.2. It can be recorded from Map 5 that there is high (more than 5%) percentage of banjar land in Jagdishpur, Baldirai, Bhadar, Lambhua, Jaisinghpur, Dhanpatganj, Kurwar, Dubepur Blocks and low percentage (less than 4%) of banjar land is seen in Gaura Jamo, Musafirkhana, Amethi, Bhnetua, Gauriganj, Kadipur, Pratappur Kamaicha, Dostpur, Akhandnagar, Kurebhar and Bhadainya Blocks of Sultanpur District.

The percentage change of banjar land from 1977-78 to 1981-82 varies from 0.96% in Pratappur Kamaicha to 41.82% in Bhadar Block which is given in Table 2.3. High percentage change (more than 12%) is observed in Bhadar, Baldirai, Jagdishpur, Gauriganj, Kurebhar, Kurwar Blocks and low percentage change (less than 12%) is recorded in Gaura Jamo, Musafirkhana, Amethi, Bhnetua, Kadipur, Lambhua, Pratappur Kamaicha, Dostpur, Akhandnagar, Jaisinghpur, Dhanpatganj, Dubepur and Bhadainya Blocks which is depicted in Map 6.

The distribution of banjar land at village level in 1981-82 (Map 9) is found in majority of the villages in which very low percentage (0 to 10%) of banjar land is observed in 2,047 villages of the district among which 667 villages are in Sadar Sultanpur, 613 in Kadipur, 405 in Amethi and 362 in Musafirkhana Tahsils. Low percentage (10% to 20%) distribution of banjar land is found in 146 villages of the district of which maximum villages are in Sadar Sultanpur (54 villages) and Kadipur Tahsils (39 villages) followed by Musafirkhana (34 villages and Amethi (29 villages) Tahsils. Medium percentage (20% to 30%) of banjar land is found in 15 villages of the district in

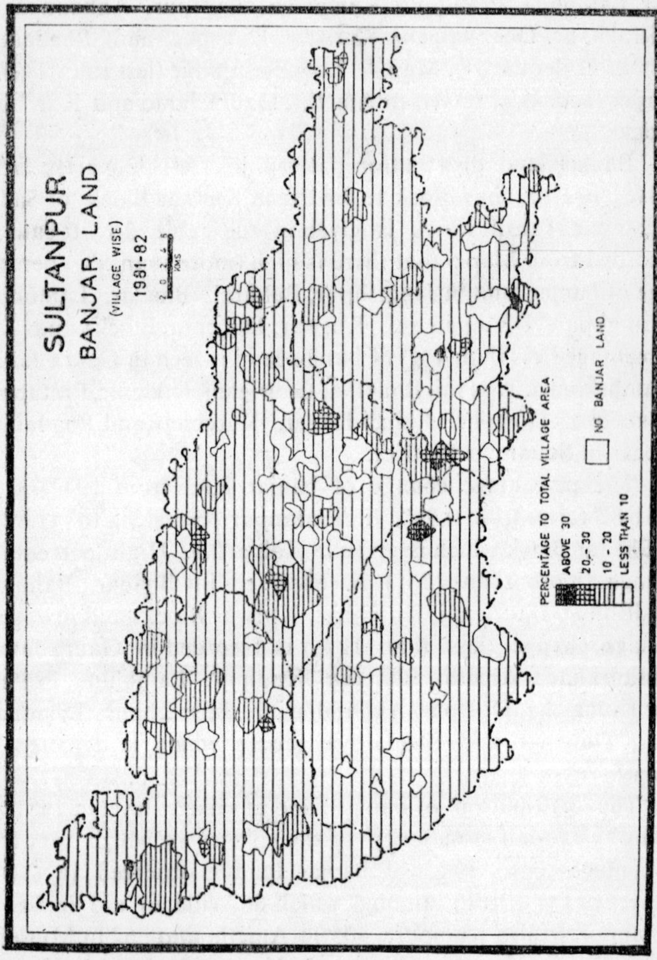
SULTANPUR
BANJAR LAND
(VILLAGE WISE)
1981-82
KMS
PERCENTAGE TO TOTAL VILLAGE AREA
ABOVE 30
20 - 30
10 - 20
LESS THAN 10
NO BANJAR LAND

MAP 9

which nine villages are in Sadar Sultanpur, five in Kadipur and one in Musafirkhana Tahsil. Medium high (30% to 40%) percentage of banjar land is observed in Piparpur village of Bhadar Block and Mugrahi, Kushela, Gulalpur villages of Pratappur Kamaich Block.

It can be observed from the above discussion that in both the periods percentage distribution of banjar land is found high (more than 3%) in Jagdishpur, Baldirai, Bhadar, Lambhua, Jaisinghpur, Dhanpatganj, Musafirkhana, Bhnetua, Kadipur, Pratappur Kamaicha, Dostpur, Akhandnagar, Kurwar, Dubepur and Bhadainya blocks. The percentage change is observed high (more than 12%) in Bhadar, Baldirai, Jagdishpur, Gauriganj, Kurwar and Kurebhar Blocks. The village level distribution of banjar land is recorded high (more than 30%) in Piparpur, Mugrahi, Kushela and Gulalpur villages of the district. Village Piparpur is further selected for analysing remedies and economics of banjar land reclamation, discussed in the following chapters.

2.3.4 Old Fallow Land

The variation in old fallow land ranges from 0.40% in Kurwar Block to 2.84% in Kadipur Block in the year 1977-78 which is depicted in Table 2.1. High percentage (more than 1.5%) of the old fallow land is evidenced in Kadipur, Pratappur Kamaicha, Dostpur, Dhanpatganj, Kurebhar Blocks which is shown in Map 4, medium percentage (1% to 1.5%) is recorded in Jagdishpur, Gaura Jamo, Musafirkhana, Bhadar, Bhnetua, Jaisinghpur, Dhanpatganj, Dubepur and Bhadainya Blocks while low percentage (less than 1%) of old fallow land is found in Baldirai, Amethi, Gauriganj, Lambhua, Akhandnagar and Kurwar Blocks.

In 1981-82 the distribution pattern of the old fallow land of the district varies from 0 48% in Kurwar Block to 2.98% in Kadipur Block as given in Table 2.2. It is recorded from the Map 5 that there is high percentage (more than 1.5%) distribution of old fallow land in Kadipur, Pratappur Kamaicha, Dostpur, Dhanpatganj, Kurebhar, Bhadainya Blocks and medium percentage (1% to 1.5%) is seen in Jaisinghpur, Gaura Jamo, Musafirkhana, Bhadar, Bhnetua, Lambhua,

Dubepur Blocks while low percentage (less than 1%) of old fallow land is found in Baldirai, Amethi, Gauriganj, Akhandnagar and Kurwar Blocks.

The percentage change of the old fallow land from 1977-78 to 1981-82 varies from 4.23% in Bhadar Block to 22.78% in Bhadainya Block as given in Table 2.3. It is clear from Map 6 that there is high percentage change of the old fallow land (more than 15%) in Jagdishpur, Baldirai, Bhnetua, Gauriganj, Akhandnagar, Kurwar, Dubepur, Bhadainya Blocks and medium percentage change (10% to 15%) is observed in Gaura Jamo, Musafirkhana, Amethi, Lambhua, Dostpur, Dhanpatganj Blocks and low percentage (less than 10%) growth of waste lands is found in Bhadar, Kadipur, Pratappur Kamaicha, Jaisinghpur and Kurebhar Blocks of the district. Thus it is clear from the above discussion that there is a significant positive change of old fallow land over a period of five years.

At the village level the distribution of old fallow land in 1981-82 is recorded to be very low (less than 10%) in 1,017 villages of the district in which Sadar Sultanpur Tahsil contributes 277 villages, Kadipur 375 villages, Amethi 203 villages and Musafirkhana 182 villages (Map 10). Low percentage (10% to 20%) distribution of old fallow land is observed in 61 villages of the district of which 27 villages are in Sadar Sultanpur, 22 villages in Kadipur, followed by 7 villages in Amethi and 5 villages in Musafirkhana Tahsil. Medium percentage (20% to 30%) of old fallow land is seen in 6 villages of Kadipur, 3 villages of Sadar Sultanpur and one village of Musafirkhana Tahsil. Medium high (30% to 40%) percentage of old fallow land is observed in Raghaipur Richhaura and Mahmudpur villages of the district, while high percentage (40% to 50%) of old fallow land is noticed in Raibigo, Bhojpur and Bhagupur villages of the study area.

Thus high percentage distribution of old fallow land is seen in Kadipur, Pratappur Kamaicha, Dostpur, Dhanpatganj, Kurebhar, Bhadainya Blocks while high percentage growth of old fallow land is observed in Jagdishpur, Baldirai, Bhnetua, Gauriganj, Akhandnagar, Kurwar, Dubepur and Bhadainya Blocks. High percentage distribution of old fallow land is recorded in Raibigo, Bhagupur and Bhojpur villages. Village Raibigo has been selected for analysing the remedies and economics of old

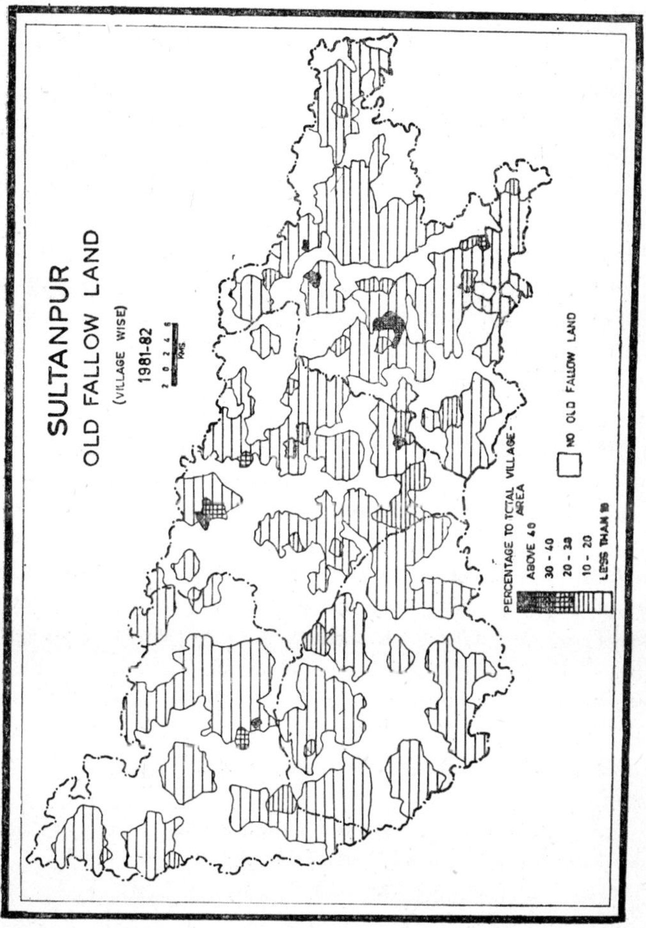
SULTANPUR
OLD FALLOW LAND
(VILLAGE WISE)
1981-82
PERCENTAGE TO TOTAL VILLAGE AREA
ABOVE 40
30 - 40
20 - 30
10 - 20
LESS THAN 10
NO OLD FALLOW LAND

MAP 10

fallow land reclamation as it has highest percentage of old fallow land to the total area of the village than any other village of the district. It has been recorded that there is no regular pattern in the distribution of old fallow land.

2.3.5 Fallow Land

It can be observed from Table 2.1 that the general distribution pattern of fallow land in 1977-78 varies from 0.30% in Jagdishpur Block to 5.59% in Bhadainya Block. High percentage of fallow land (more than 4%) is found in Lambhua and Bhadainya Blocks, Medium percentage (2% to 4%) of fallow land is recorded in Gaura Jamo, Amethi, Bhadar, Bhnetua, Gauriganj, Kadipur, Dostpur, Jaisinghpur, Dhanpatganj, Kurebhar and Dubepur Blocks which is depicted in the Map 4. Low percentage (less than 2%) of fallow land is seen in Jagdishpur, Musafirkhana, Baldirai, Pratappur Kamaicha, Akhandnagar and Kurwar Blocks.

The general distribution pattern of fallow land in 1981-82 which is given in the Table 2.2 ranges from 0.51% in Jagdishpur Block to 5.76% in Bhadainya Block. It is clear from Map 5 that high percentage (more than 4%) distribution of fallow land is seen in Lambhua and Bhadainya Blocks while medium (2% to 4%) percentage of fallow land is observed in Gaura Jamo, Jagdishpur, Amethi, Bhadar, Bhnetua, Gauriganj, Kadipur, Dostpur, Pratappur Kamaicha, Jaisinghpur, Dhanpatganj, Kurebhar, Dubepur Blocks. Lastly, low percentage (less than 2%) of fallow land is recorded in Jagdishpur, Musafirkhana, Pratappur Kamaicha, Akhandnagar and Kurwar Blocks.

The percentage growth of fallow land from 1977-78 to 1981-82 (Table 2.3) varies from 2.95% in Bhadainya Block to 39.32% in Jagdishpur Block. It is seen from Map 6 that high percentage (more than 15%) growth in fallow land is found in Jagdishpur, Pratappur Kamaicha and Kurwar Blocks. Medium percentage (8% to 15%) change is observed in Gaura Jamo, Musafirkhana, Baldirai, Amethi, Bhnetua, Akhandnagar, Dhanpatganj Blocks and low percentage growth (less than 8%) of fallow land is recorded in Bhadar. Gauriganj, Kadipur, Lambhua, Dostpur, Jainsinghpur, Kurebhar, Dubepur and

Bhadainya Blocks which shows that there is a significant change in fallow land at block level in Sultanpur District.

It can be seen from Map 11 that the distribution pattern of fallow land in 1981-82 at village level is observed to be very low (less than 10%) in 1,169 villages of the district of which 386 villages are in Sadar Sultanpur followed by 297 villages in Amethi, 279 in Kadipur and 207 villages in Musafirkhana Tahsils. Low percentage (10% to 20%) of fallow land is found in 45 villages of Sadar Sultanpur and Kadipur Tahsils, 27 villages in Amethi Tahsil and 8 villages in Musafirkhana Tahsil. Medium percentage (20% to 30%) of fallow land is observed in 14 villages of Kadipur Tahsil, 7 villages of Sadar Sultanpur, 6 villages of Amethi and one village of Musafirkhana Tahsil. Medium high percentage (30% to 40%) of fallow land is seen in 14 villages of Kadipur, 7 villages of Sadar Sultanpur, 6 villages of Amethi and one village of Musafirkhana Tahsils. Medium high percentage (30% to 40%) of fallow land is observed in Ahibaranpur, Rampur Jaria, Devaritaranpatti villages of Kadipur Tahsil, Partos, Jagdishpur and Pitambarpur villages of Amethi Tahsil and Bhadainya, Bharopur and Sewara villages of the Sadar Sultanpur Tahsil.

Thus there is high percentage distribution of fallow land in Lambhua and Bhadainya Blocks while percentage change is observed high in Jagdishpur, Pratappur Kamaicha and Kurwar Blocks. Medium high percentage distribution of fallow land is seen in Ahibaranpur, Rampur Jaria, Devaritaranpatti, Partos Jagdishpur, Pitambarpur, Bhadainya, Bharopur and Sewara villages of the district. It can be recorded from Map 6 that there is a little regularity in the distribution of fallow land at village level in Sultanpur District. Village Bhadainya of Sadar Sultanpur Tahsil has been selected for analysing the remedies and economics of fallow land reclamation.

2.3.6 Kankrili Land

The distribution pattern of the kankrili land in 1977-78 ranges from 0.01% in Pratappur Kamaicha Block to 0.27% in Kurwar Block which is given in Table 2.1. It is recorded from Map 4 that there is high percentage (more than 0.10%) of kankrili land in Jaisinghpur, Dhanpatganj, Kurwar, Kurebhar Blocks

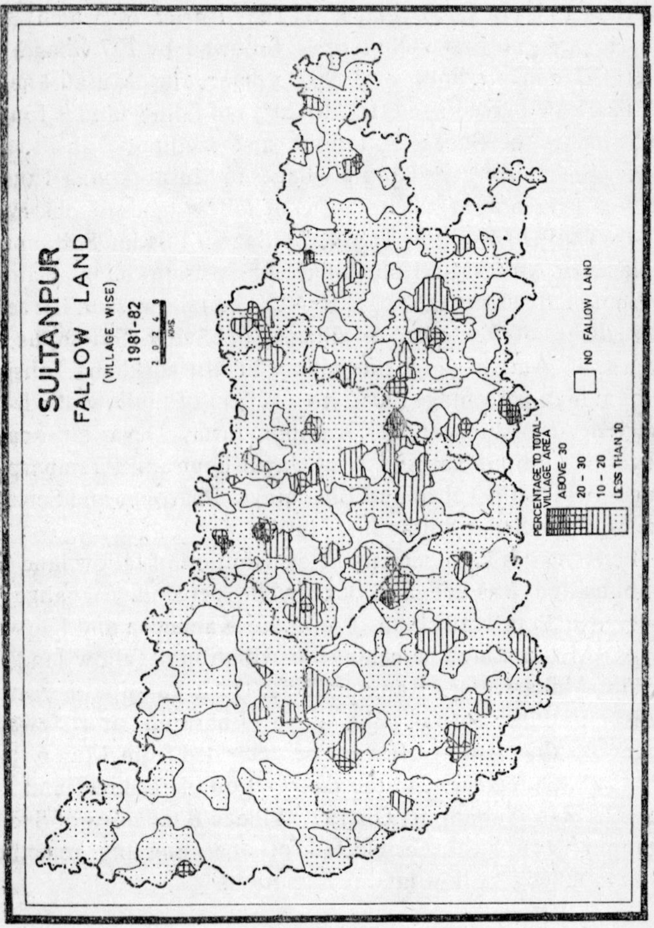
SULTANPUR
FALLOW LAND
(VILLAGE WISE)
1981-82
KMS
PERCENTAGE TO TOTAL VILLAGE AREA
ABOVE 30
20 - 30
10 - 20
LESS THAN 10
NO FALLOW LAND

MAP 11

and low percentage (less than 0.10%) of kankrili land is observed in Kadipur, Pratappur Kamaicha, Dostpur, Dubepur and Bhadainya Blocks.

The high percentage (more than 0.10%) distribution of kankrili land in 1981-82 is observed in Jaisinghpur, Dhanpatganj, Kurebhar, Kurwar, Dubepur and Bhadainya Blocks of Sadar Sultanpur Tahsil which is given in Table 2.2, while percentage distribution of kankrili land ranges from 0.02% in Pratappur Kamaicha to 0.33% in Kurwar Block. Low percentage (less than 0.10%) distribution of kankrili land is found in Pratappur Kamaicha, Kadipur and Dostpur Blocks of Kadipur Tahsil which is depicted in Map 5.

The percentage growth of kankrili land from 1977-78 to 1981-82 ranges from 1.11% in Jaisinghpur to 42.85% in Pratappur Kamaicha Blocks as given in Table 2.3. It is clear from Map 6 that high percentage (more than 25%) growth of kankrili land is recorded in Kadipur, Jaisinghpur, Dhanpatganj, Kurebhar and Dubepur Blocks.

The distribution of kankrili land in 1981-82 at village level is recorded in 51 villages of the district (Map 12) of which 41 villages are in Sadar Sultanpur Tahsil and 10 villages are in Kadipur Tahsil having very low percentage range (less than 10%) of kankrili land which is depicted in Map 12. Low percentage of kankrili land (10% to 20%) is recorded in Seur Chamurkha, Karamchandpur, Ardih, Guregaon of Sadar Sultanpur Tahsil and Babarpur village of Kadipur Tahsil.

It can be seen from the above discussion that the high percentage distribution of kankrili land (more than 0.10%) is in Jaisinghpur, Dhanpatganj, Kurwar, Kurebhar, Dubepur and Bhadainya Blocks while percentage change is recorded high (more than 25%) in Pratappur Kamaicha, Dostpur, Kurwar and Bhadainya Blocks. The village level distribution of kankrili land is found more in Seurchamurkha, Karamchandpur, Ardih, Guregaon and Babarpur villages. Village Seurchamurkha, having the highest percentage of kankrili land, is selected for analysing remedies and economics of kankrili land reclamation. It has also been recorded that there is no regular pattern in the distribution of kankrili land at village level in Sultanpur District.

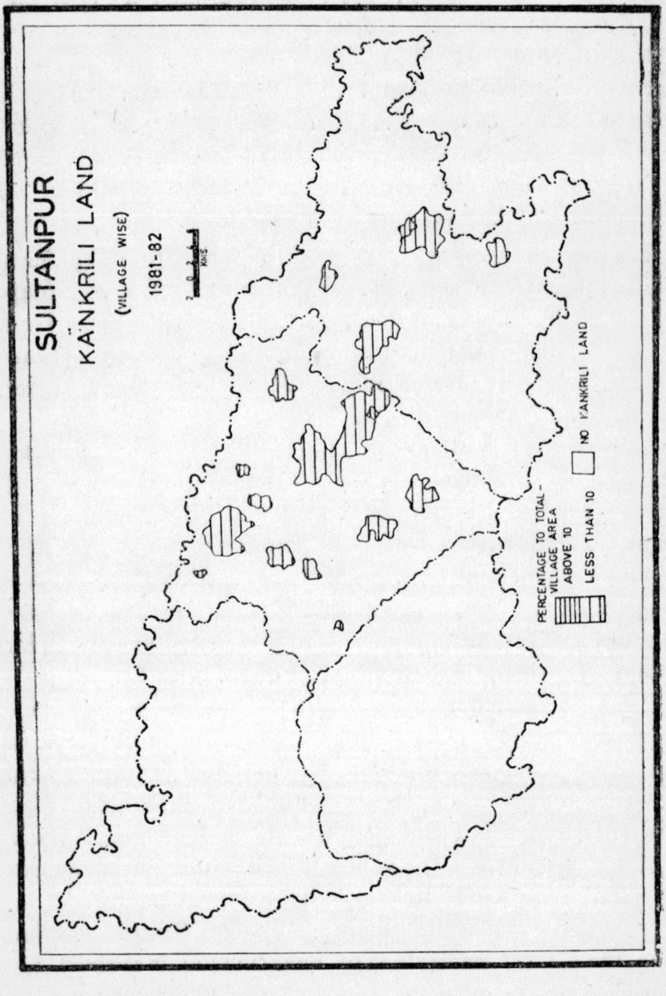
SULTANPUR
KANKRILI LAND
(VILLAGE WISE)
1981-82
PERCENTAGE TO TOTAL-VILLAGE AREA
ABOVE 10
LESS THAN 10
NO KANKRILI LAND

MAP 12

2.3.7 Ravine Land

The general distribution pattern of ravine land ranges from 0.04% in Pratappur Kamaicha to 1.66% in Baldirai Block in 1977-78 as given in Table 2.1. It can be seen in Map 4 that there is high percentage (more than 0.50%) distribution of ravine land in Jaisinghpur, Dhanpatganj, Kurebhar, Jagdishpur, Musafirkhana, Baldirai, Kurwar, and Bhadainya Blocks, while low percentage (less than 0.50%) distribution of ravine land is observed in Kadipur, Lambhua, Pratappur Kamaicha, Dostpur and Dubepur Blocks, near river Gomati.

The distribution of ravine land in 1981-82 (Table 2.2) ranges from 0.06% in Pratappur Kamaicha to 1.83% in Baldirai Block. High percentage (more than 0.50%) distribution of ravine land is recorded in Jagdishpur, Musafirkhana, Baldirai Blocks of Musafirkhana Tahsil and Jaisinghpur, Dhanpatganj, Kurebhar, Kurwar and Bhadainya Blocks of Sadar Sultanpur Tahsil which is depicted in Map 5 and low percentage (less than 0.50%) distribution of ravine land is found in Kadipur Lambhua, Pratappur Kamaicha, Dostpur Blocks of Kadipur Tahsil and Dubepur Block of Sadar Sultanpur Tahsil.

The percentage growth of ravine land from 1977-78 to 1981-82 (Table 2.3) ranges from 7.06% in Dhanpatganj Block to 38.24% in Kurebhar Block of Sadar Sultanpur Tahsil. It is clear from Map 6 that there is high percentage (more than 20%) growth of ravine land in Kadipur, Lambhua, Pratapur Kamaicha, Dostpur, Kurebhar, Kurwar and Bhadainya Blocks, while low percentage (less than 20%) change is observed in Jagdishpur, Musafirkhana, Baldirai, Jaisinghpur, Dhanpatganj and Dubepur Blocks.

The village level distribution pattern of ravine land in 1981-82 which is depicted in Map 13 is found very low (less than 10%) in 180 villages of the district in which 66 villages are in Sadar Sultanpur, 38 villages in Kadipur and 76 villages in Musafirkhana Tahsil. Low percentage (10% to 20%) distribution of ravine land is observed in 10 villages of the district of which 6 villages are in Sadar Sultanpur and 4 villages in Musafirkhana Tahsil. Medium percentage (20% to 30%) of ravine land is found in Khajapur and Upadhyapur villages of Sadar Sultanpur Tahsil and medium high (30% to 40%) percentage of

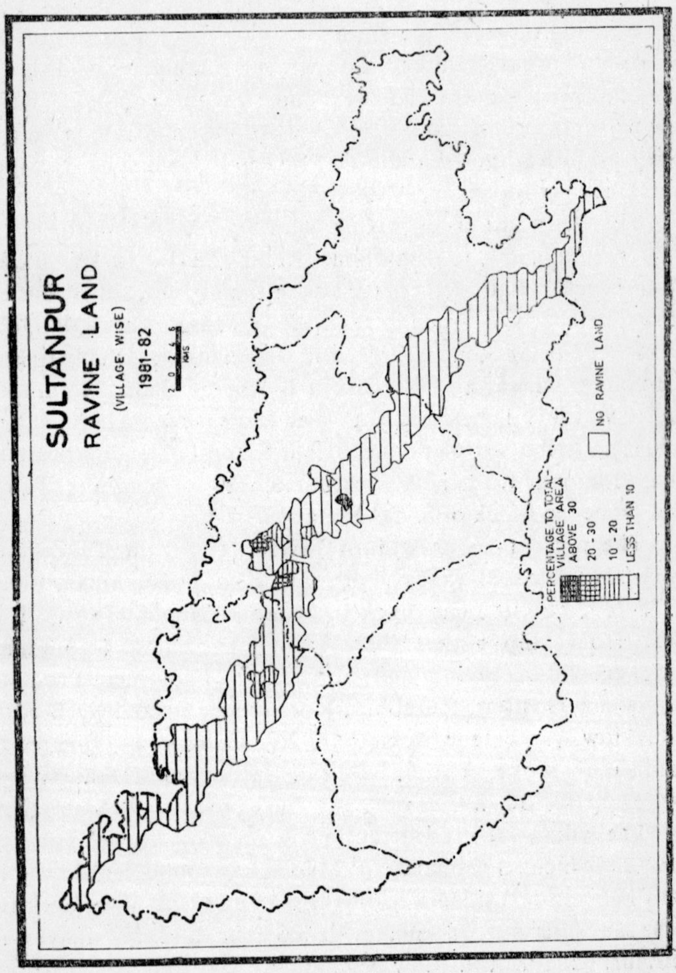
SULTANPUR
RAVINE LAND
(VILLAGE WISE)
1981-82
KMS
PERCENTAGE TO TOTAL VILLAGE AREA
ABOVE 30
20 - 30
10 - 20
LESS THAN 10
NO RAVINE LAND

MAP 13

ravine land is recorded in Ratanpur village of Sadar Sultanpur Tahsil.

On the whole a high percentage (more than 0.50%) distribution of ravine land is recorded in Jagdishpur, Musafirkhana, Baldirai, Jaisinghpur, Dhanpatganj, Kurebhar, Kurwar and Bhadainya Blocks while high percentage change (more than 20%) of ravine land is recorded in Kadipur, Lambhua, Pratappur Kamaicha, Dostpur, Kurebhar, Kurwar and Bhadainya Blocks. The village level distribution of ravine land is recorded high (more than 20%) in Ratanpur, Upadhyapur and Khajapur villages of Sadar Sultanpur Tahsil. Village Ratanpur (having the highest percentage or ravine land) is selected for analysing remedies and economics of ravine land reclamation. There is a continuity in the distribution pattern of ravine land as it is mainly confined to river Gomati, lying on either side of it.

2.3.8 Other Types of Waste Land

The distribution of other types of waste land in 1977-78 ranges from 0.63% in Musafirkhana to 8.01% in Akhandnagar Block which is given in the Table 2.1. Map 4 shows that there is high percentage (more than 3%) distribution of other types of waste land in Bhnetua, Gauriganj, Kadipur, Lambhua, Akhandnagar, Jaisinghpur, Kurebhar and Bhadainya Blocks while medium percentage (2% to 3%) is found in Amethi, Bhadar, Pratappur Kamaicha, Dhanpatganj Blocks and low percentage (less than 2%) of other types of waste land is observed in Jagdishpur, Gaura Jamo, Musafirkhana, Baldirai, Kurwar and Dubepur Blocks.

The percentage distribution of other types of waste land in 1981-82 (Table 2.2) ranges from 0.83% in Musafirkhana to 9.26% in Akhandnagar Block. High percentage (more than 4%) of other types of waste land is recorded in Bhnetua, Kadipur, Lambhua, Dostpur Blocks which is depicted in Map 5, while medium percentage (2% to 4%) is recorded in Gaura Jamo, Amethi, Bhadar, Gauriganj, Jaisinghpur, Kurebhar, Bhadainya Blocks and low percentage (less than 2%) of other types of waste land is found in Jagdishpur, Musafirkhana, Baldirai, Kurwar and Dubepur Blocks.

The percentage growth of other types of waste land from

1977-78 to 1981-82 (Table 2.3) ranges from 4.54% in Dostpur Block to 29.93% in Baldirai Block. It can be observed from Map 6 that there is high percentage growth (more than 15%) of other types of waste land in Baldirai, Akhandnagar, Musafirkhana, Kurwar Blocks while medium percentage (5% to 15%) change is observed in Jagdishpur, Gaura Jamo, Amethi, Bhadar, Bhnetua, Kadipur, Lambhua, Dubepur Blocks and low percentage (less than 5%) growth of other types of waste land is found in Bhnetua, Pratappur Kamaicha, Dostpur, Dhanpatganj, Kurebhar and Bhadainya Blocks.

It can be seen from Map 14 that the percentage distribution of other types of waste land in 1981-82 at village level is very low (less than 10%) in 1,756 villages of the district of which 551 villages are recorded in Sadar Sultanpur, 527 villages in Kadipur followed by 403 villages in Amethi and 275 villages in Musafirkhana Tahsils. Low percentage (10% to 20%) distribution of other types of waste land is observed in 138 villages of the district out of which 33 villages are in Sadar Sultanpur, 88 villages in Kadipur, 11 villages in Amethi and 6 villages in Musafirkhana Tahsils. Medium percentage (20% to 30%) distribution of other types of waste land is found in 22 villages of the district of which 17 villages are in Kadipur, 4 villages in Sadar Sultanpur and one village in Musafirkhana Tahsils. Medium high (30% to 40%) percentage of other types of waste land is observed in Dadara village of Musafirkhana Tahsil and Kundabharo village of Kadipur Tahsil.

Thus high percentage (more than 4%) distribution of other types of waste land is observed in Bhnetua, Kadipur, Lambhua, Dostpur Blocks while percentage change is recorded high (more than 15%) in Baldirai, Akhandnagar, Musafirkhana and Kurwar Blocks. The village level distribution of other types of waste land is observed to be high (more than 30%) in Dadara and Kundabharopur villages of the district. Village Dadara having the highest percentage of other types of waste land was selected for analysing the remedies and economics of reclamation. It has been recorded that there is a marked regularity in the distribution pattern of other types of waste land at village level.

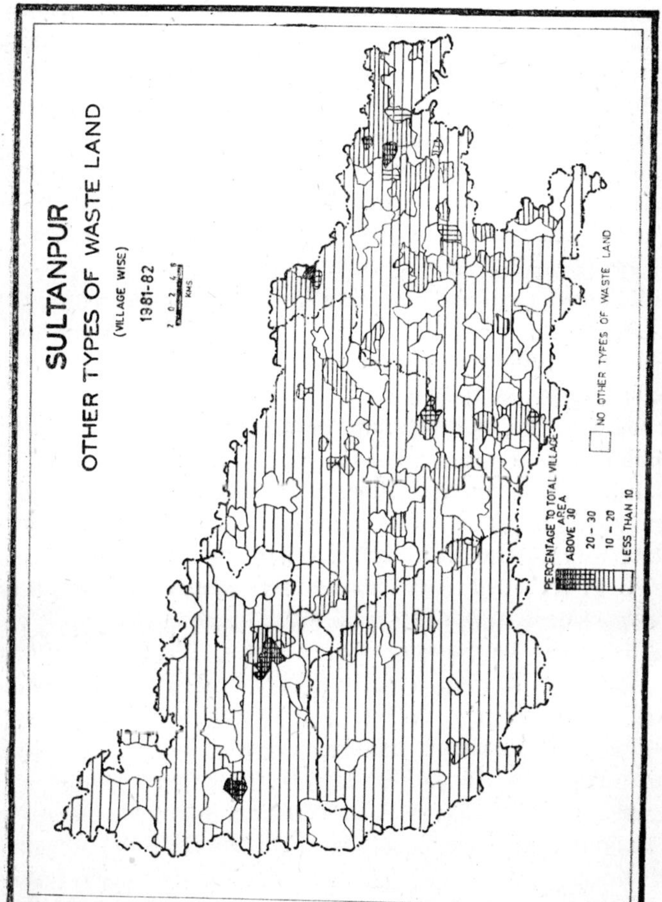
SULTANPUR
OTHER TYPES OF WASTE LAND
(VILLAGE WISE)
1981-82
KMS
PERCENTAGE TO TOTAL VILLAGE AREA
ABOVE 30
20 - 30
10 - 20
LESS THAN 10
NO OTHER TYPES OF WASTE LAND

MAP 14

2.3.9 Waste Lands

The general distribution of waste lands of Sultanpur District in 1977-78 to 1981-82 (Table 2.1) ranges from 8.21% in Musafirkhana Block to 16.93% in Lambhua Block. It is observed from Map 4 that there is high percentage (more than 13%) distribution of waste lands in Bhadar, Kadipur, Lambhua, Dostpur, Akhandnagar, Kurebhar and Bhadainya Blocks. Medium percentage (10% to 13%) of waste lands is observed in Gaura Jamo, Bhnetua, Jaisinghpur, Dhanpatganj, Dubepur Blocks while low percentage (less than 10%) distribution of waste lands is found in Jagdishpur, Musafirkhana, Baldirai, Amethi, Gauriganj, Pratappur Kamaicha and Kurwar Blocks.

In 1981-82 (Table 2.2) waste lands distribution varies from 9.14% in Musafirkhana to 17.69% in Lambhua Block. High percentage (more than 13%) distribution of waste lands (Map 5) is observed in Bhadar, Lambhua, Kadipur, Dostpur, Akhandnagar, Jaisinghpur, Dhanpatganj, Kurebhar and Bhadainya Blocks. Medium percentage (10% to 13%) of waste lands is recorded in Jagdishpur, Gaura Jamo, Baldirai, Gauriganj, Bhnetua and Dubepur Blocks while low percentage (less than 10%) of waste lands is observed in Musafirkhana, Amethi, Pratappur Kamaicha and Kurwar Blocks.

It is recorded from Table 2.3 that there is high percentage growth (more than 13%) of waste lands from 1977-78 to 1981-82 in Jagdishpur, Gaura Jamo, Bhadar, Gauriganj, Dostpur and Kurwar Blocks which is depicted in Map 6. Medium percentage change (8% to 13%) of waste lands is found in Musafirkhana, Baldirai, Amethi, Kadipur, Dubepur and Bhadainya Blocks while low percentage (less than 8%) growth of waste lands is observed in Bhnetua, Lambhua, Pratappur Kamaicha, Jaisinghpur, Dhanpatganj and Kurebhar Blocks.

The general distribution of waste lands in 1981-82 at village level is recorded in 2,494 villages of the district (Map 15) which constitute about 98% of the total villages of the district. Very low percentage (less than 10%) of waste lands is found in 424 villages of Sadar Sultanpur, 356 villages in Kadipur, 285 villages in Amethi and 253 villages in Musafirkhana Tahsils. Low percentage (10% to 20%) of waste lands is observed in

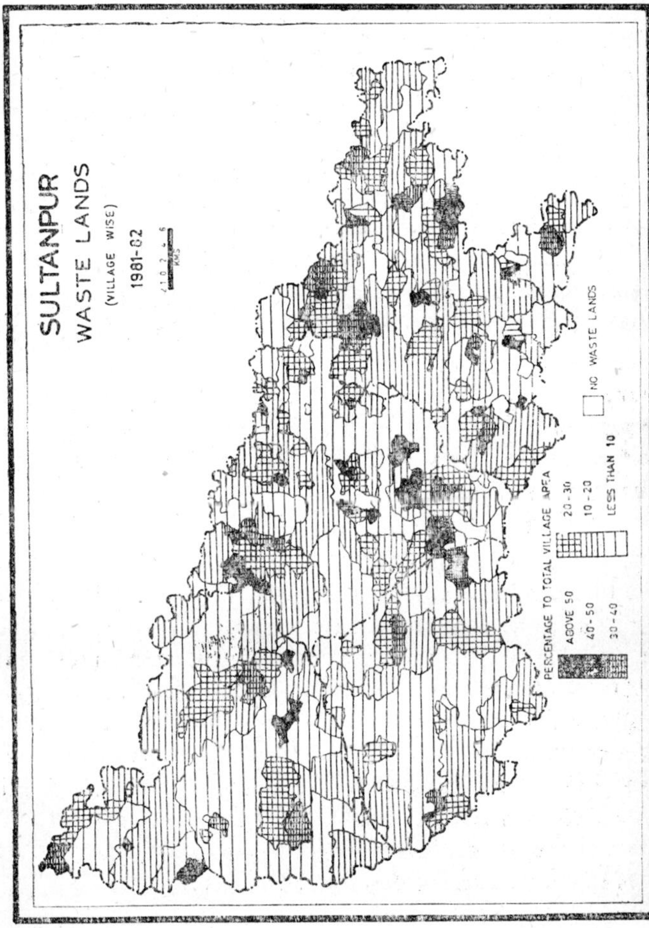
SULTANPUR
WASTE LANDS
(VILLAGE WISE)
1981-82
KMS
PERCENTAGE TO TOTAL VILLAGE AREA
ABOVE 50
40-50
30-40
20-30
10-20
LESS THAN 10
NO WASTE LANDS

MAP 15

798 villages of the district where a maximum of 267 villages are in Sadar Sultanpur, 259 villages in Kadipur, 148 villages in Amethi and 124 villages in Musafirkhana Tahsils. Medium percentage (20% to 30%) of waste lands is found in 248 villages of the district of which maximum villages are in Kadipur (103 villages), 83 villages in Sadar Sultanpur, 33 villages in Amethi and 29 villages in Musafirkhana Tahsils which is depicted in Map 15. Medium high (30% to 40%) percentage of waste lands is observed in 89 villages of the district of which 34 (maximum) villages are in Kadipur, 33 villages in Sadar Sultanpur, 12 villages in Amethi and 10 villages in Musafirkhana Tahsils. It is recorded that there is high percentage (40% to 50%) distribution of waste lands in 28 villages of the district of which 13 (maximum) villages are in Sadar Sultanpur, 9 villages in Kadipur, 4 villages in Amethi and 2 villages in Musafirkhana Tahsils. Very high (more than 50%) percentage of waste lands is recorded in 13 villages of the district of which 7 (maximum) villages are in Kadipur, 4 villages in Sadar Sultanpur and 2 villages in Musafirkhana Tahsils.

It can be seen from the above discussion that the high percentage (more than 13%) distribution of waste lands is found in Bhadar, Kadipur, Lambhua, Dostpur, Akhandnagar, Kurebhar and Bhadainya Blocks while percentage change over a five year period is observed to be high (more than 13%) in Jagdishpur, Gaura, Jamo Bhadar, Gauriganj, Dostpur aud Kurwar Blocks. Very high percentage (more than 50%) distribution of waste lands at village level is observed in 7 villages of Sadar Sultanpur, 4 villages of Kadipur and 2 villages of Musafirkhana Tahsils. It has also been observed that there is a regularity in the distribution of waste lands at village level of Sultanpur District.

2.4 CONCLUSION

Waste lands of Sultanpur District are divided into 8 types: waterlogged land, usar land, banjar land, old fallow land, fallow land, kankrili land, ravine land and other types of waste land. It was indicated earlier that the area under waste lands in Sultanpur was quite high (12.85%) in 1977-78 and has increased to 13.72% in 1981-82. The distribution of waste lands, according to categories, indicates that banjar land forms a substantially high proportion whereas kankrili land is only nominal being less than 0.1%. The change in waste land over the five-year period is recorded as 10.93% while, category wise, maximun change is observed in waterlogged land (29.96% and minimum change is observed in fallow land (7.92%). It has been discussed that there is a wide variation in the distribution of waste lands within the district and also at tahsil level. The variations have been depicted for the years 1977-78 and 1981-82 and the change over the five year period is also discussed. But the categorywise village level distribution of waste lands is discussed only for the year 1981-82.

It is observed that high percentage of waterlogged land (more than 1%) is found in Gaura Jamo, Dostpur and Bhadainya Blocks, while the change from 1977-78 to 1981-82 is observed to be high (more than 50%) in Kadipur and Dostpur Blocks and the negative change (—30.76%) in waterlogged land is found in Gauriganj Block where the problem has been tackled. Village level distribution of waterlogged land is seen to be high (more than 20%) in Malapur, Jagdishpur and Asrawan villages of the district.

A high percentage of usar land (more than one) is seen in Bhadar, Lambhua, Dostpur and Bhadainya Blocks and percentage change (more than 30) is found to be high in Jagdishpur, Mesafirkhana, Bhadar, Gauriganj, Lambhua, Akhandnagar, Dhanpatganj and Kurwar Blocks. Also a high percentage (more than 30) distribution of usar land at village level is observed in Ghazipur, Bhainsada, Dilipgarh and Purainachandrabhan villages of the district.

The distribution of banjar land is found to be high (more than 3%) in Jagdishpur, Baldirai, Bhadar Lambhua, Jaisinghpur, Dhanpatganj, Musafirkhana, Bhnetua, Kadipur, Pratappur, Kamaicha, Dostpur, Akhandnagar, Kurwar, Dubepur and Bhadainya Blocks. The village level distribution of banjar land is recorded high (more than 30%) in Piparpur, Mugrahi, Kushela and Gulalpur villages of the district.

A high percentage distribution of old fallow land (more than 1.5%) is observed in Kadipur, Pratappur Kamaicha, Dostpur, Dhanpatganj and Kurebhar Blocks and percentage change is seen to be high (more than 15) in Jagdishpur, Baldirai, Bhnetua, Gauriganj, Akhandnagar, Kurwar, Dubepur and Bhadainya Blocks. The distribution of old fallow land at village level is high (more than 40%) in Raibigo, Bhagupur and Bhojpur villages.

There is high percentage (more than 4) distribution of fallow land in Lambhua and Bhadainya Blocks while percentage growth of fallow land is observed to be high (more than 15) in Jagdishpur, Pratappur Kamaicha and Kurwar Blocks. High percentage (more than 30) distribution of fallow land is seen in Ahibaranpur, Rampur, Jaria, Devantaranpatti, Partos, Jagdishpur, Pitambarpur, Bhadainya, Bharopur and Sewara villages of the district.

High percentage distribution of kankrili land (more than 0.1) is seen in Jagdishpur, Dhanpatganj, Kurewar, Kurebhar, Dubepur and Bhadainya Blocks while high percentage change (more than 25) is recorded in Pratappur Kamaicha, Dostpur, Kurwar and Bhadainya Blocks. The village level distribution of kankrili land is found to be high (more than 10%) in Seurchamurkha, Karamchandpur, Ardih, Guregaon and Babarpur villages of the district.

The distribution of ravine land is observed to be high (more than 0.50) in Jagdishpur, Musafirkhana, Baldirai, Jaisinghpur, Dhanpatganj, Kurebhar, Kurwar and Bhadainya Blocks and a high percentage change (more than 20) is observed in Kadipur, Lambhua, Pratappur Kamaicha, Dostpur, Kurebhar, Kurwar and Bhadainya Blocks. The village level distribution of ravine land is found high (more than 20%) in Ratanpur, Khajapur and Upadhyapur villages.

A high percentage distribution of other types of waste land (more than 4) is recorded in Bhanetua, Kadipur, Lambhua and Dostpur Blocks and percentage growth of other types of waste land is found to be high (more than 15) in Baldirai, Akhandnagar, Musafirkana and Kurwar Blocks. The village level distribution of other types of waste land is observed to be high (more than 30%) in Dadara and Kundabharopur village of the district.

The general distribution of waste lands is found to be high (more than 13%) in Bhadar, Kadipur, Lambhua, Dostpur, Akhandnagar, Kurebhar and Bhadainya Blocks while percentage change from 1977-78 to 1981-82 is high (more than 13) in Jagdishpur, Gaura Jamo, Bhadar, Gauriganj, Dostpur and Kurwar Blocks. A very high percentage (more than 50) distribution of waste lands is found at village level in 7 villages of the Sadar Sultanpur, 4 villages of Kadipur and 2 villages of Musafirkhana Tahsils. One can witness a continuity in the distribution pattern of waste lands at village level. From the above, one finds that about 98% villages of the district suffer from the problem of waste lands in general. However more than half the villages (52%) are at a low level of less than 10% of their total area covered with waste lands.

3

FACTORS IN WASTE LANDS

3.1 INTRODUCTION

The preceding chapter was devoted to a discussion of typology and distribution of waste lands at tahsil, block and village level of Sultanpur District. The spatial distribution and the marked definite growth of the waste lands at the various levels of the district provide an incentive to probe into the cause-and-effect phenomenon of the waste lands. The causative factors responsible for the development of waste lands are numerous but the predominant ones can easily be identified and analysed. Waste lands as a type of land form can be attributed to complex processes interacting at varying intensities.

Thus it becomes necessary to investigate the factors influencing waste lands of Sultanpur District. For the primary investigation of the factors determining waste lands of the district it is assumed that the development of waste lands occurs through an unhealthy interaction of three forces, viz. man, nature and technology. In other words when the environment is severe, man's ability to interact with nature to his advantage is curtailed, and he fails to fully avail himself of the land resources. Again when the level of technology is very low, the waste land increases. But, on the whole, the combined impact of man and nature interaction and the varying levels of technological inputs are considered to be the main factors in the formation of waste lands. Though the natural factors assume primary importance, man's varying levels of ability to utilise land with the help of technological innovations also accelerate waste land formation. Therefore, all the responsible factors for waste land formation must be clearly identified, and investigated into, otherwise first remedial measures would be impossible.

On the basis of the above assumption various environmental (morphometric, climatic, hydrologic) and anthropogenic factors have been selected for analysing each waste land category separately, as generalisation cannot be made for all the waste land types together, because of the varying impact of the factors on varying waste land types.

Glancing back at the past, very little attention has been paid to the factors influencing formation of waste lands but for a few types (usar, waterlogged, ravine) which have been surveyed and studied by the Planning Commission and Ministry of Agriculture, Indian Agricultural Research Institute and Central Soil Salinity Research Institute, Karnal, laid a special emphasis on the pedogeny of the saline and alkali land associated with their distribution and reclamation.

In this chapter an attempt has been made to analyse various variables of environmental and anthropogenic factors that influence different types of waste land. These factors have been grouped under two main heads, environmental (morphometric, climatic, hydrologic, pedogenic) and anthropogenic, with subdivisions as given below:

Environmental Factors

(a) Morphometric

Slope
Ruggedness Number

(b) Climatic

Mean Annual Rainfall

(c) Hydrologic

Drainage Density
Quality of water in *p*H
Behaviour of water Table

(d) Pedogenic

Nitrogen of the Soil

Phosphate of the Soil
Potash of the Soil
*p*H value of the Soil
Thickness of the Top Soil
Depth of the Base Rocks.

Anthropogenic Factors

Distance of the Canal from Field
Land Concentration (Gini's coefficient)

Since it would have been an impossible task to discuss the factors for each type of waste land for all the villages of the district, the analysis has been carried out for 15 villages selected on the basis of quartiling.

3.2 FACTORS IN WATERLOGGED LAND

Waterlogging occurs due to rise in the sub-soil water table which happens mostly in overirrigated areas. This high sub-soil water table is found due to the seepage of water from unlined canal banks or from the encroachment of flood zones. These excessive irrigation water seeps down and adds to the sub-soil water table as the outflow does not increase correspondingly with the rise in sub-soil water.

To analyse the factors influencing water logged lands of Sultanpur District, 15 villages have been selected on the basis of quartiling from all the waterlogged villages of the district. This has been recorded from Appendix 1 that the selected very high percentage (19.25 to 21.94) waterlogged land villages are Malapur Jagdishpur, Asrawan and Katawana and high percentage (15.61 to 17.14) villages are Kaman Girpur, Kurapur and Haripur. Medium percentage (10.28 to 14.53) waterlogged land sample villages are Nathupur, Baijipur and Bokhrepur and low percentage (4.49 to 8.61) sample villages are Tenduakaji, Upadhyapur, Bhatpura and Mebupur Perchhauli while very low percentage (0.25 to 3.20) selected waterlogged land sample villages are Mustafabad Saraiya, Kadipur Khurd and Haripur. For these villages causative factors have been identified under the following heads.

3.2.1 Morphometric (Slope X_1 and Ruggedness Number X_2)

Waterlogging as observed in the sample villages is mainly confined to natural low-lying areas where there is every possibility of water accumulation through seepage and more inflow of water either through sub-surface drainage or due to overirrigation. Slope seems to be a predominant parameter in determining the waterlogging condition of the district as all the waterlogging patches are confined to low-lying areas. Slope is found very high (2.67° to 3.10°) in Malapur Jagdishpur, Asrawan and Katawana villages and very low (0.85° to 1.38°) in Mustafabad Saraiya, Kadipur Khurd and Haripur villages (Table 3.1).

The distribution pattern of slope and waterlogged land in the sample villages proves that there is a positive relationship between these two variables (Table 3.2) which is significant at 1% level of significance.

Very high ruggedness number (0.28 to 0.32) is recorded in those villages where the percentage of waterlogging is also very high, i.e. in Malapur Jagdishpur, Asrawan and Katawana villages while very low ruggedness number is observed in those villages where percentage of waterlogged land is also very low (Table 3.1). Ruggedness number can also be a determinant in enhancing waterlogging condition. The more rugged the land, the more is the possibility of waterlogging conditions, as both surface and sub-surface conditions will be effected by the variations in the nature of the topography. Areas with high altitudes do account for greater slope amounts and ruggedness number. Hence these variables are correlated among themselves and provide greater possibility of water accumulation.

The result of the correlation matrix, as given in Table 3.2, proves the assumption that there is a positive significant correlation between ruggedness number X_2 (independent variable) and waterlogged land Y (dependent variable) which is significant at 1% level of significance because a high ruggedness number provides positive condition for more surface water flow and collection of water in depressed areas, which is helpful in raising the water table.

TABLE 3.1 : **Factors in Waterlogged Land, 1981-82**

Sample Villages	% of water logged land to total area (Y)	Slope in degrees (X_1)	Ruggedness number (X_2)	Mean Annual Rainfall in mms. (X_3)	Drainage Density (X_4)	Quality of water in pH (X_5)	Behaviour of water table in metres (X_6)	Pedogenic: Nitrogen (X_7)	Pedogenic: Phosphate (X_8)	Pedogenic: Potash (X_9)	Pedogenic: pH Value (X_{10})	Distance of the Canal from field in metres (X_{11})
1	2	3	4	5	6	7	8	9	10	11	12	13
Malapur												
Jagdishpur	21.94	3.10	0.32	219	8.7	7.63	1.5	1.04	1.58	3.96	7.66	46
Asrawan	20.05	2.94	0.31	215	8.4	7.66	1.8	1.03	1.54	3.87	7.53	52
Katawana	19.25	2.67	0.28	213	8.1	7.52	1.3	1.01	1.51	3.78	7.67	59
Kamangirpur	17.14	2.54	0.27	214	7.9	7.55	2.1	1.07	1.56	3.68	7.45	68
Kamapur	16.15	2.48	0.25	210	7.5	7.43	2.6	1.09	1.58	3.74	7.49	70
Haripur	15.61	2.65	0.28	218	7.2	7.46	2.9	2.00	1.59	3.65	7.52	75
Nathupur	14.53	2.38	0.22	198	6.8	7.38	3.2	2.03	1.63	3.61	7.64	78
Baijipur	12.58	2.10	0.20	212	6.5	7.30	3.7	2.05	1.65	3.58	7.75	83
Bokhripur	10.28	1.85	0.18	218	6.1	7.21	3.8	2.07	1.68	3.53	7.56	87

Tajuddinpur	8.61	1.68	0.16	220	5.8	7.24	4.3	2.08	1.72	3.51	7.58	90
Ugharpur	5.90	1.56	0.13	216	5.3	7.37	4.6	2.10	1.82	3.48	7.61	93
Maharani Pachhim	4.49	1.43	0.10	213	4.6	7.45	4.9	2.12	1.89	3.32	7.73	95
Mustafabad Saraiya	3.20	1.38	0.08	205	3.2	7.52	5.2	2.13	1,92	3.24	7.45	100
Kadipur Khurd	1.31	1.05	0.06	209	2.7	7.61	5.6	2.16	1.94	3.11	7.57	105
Haripur	0.25	0.85	0.04	203	1.3	7.67	5.8	2.18	1.93	2.98	7.79	115

TABLE 3.2 : Correlation Matrix

Sl. No.	Variables	Y	X_1	X_2	X_3	X_4	X_5
Y	% of Waterlogged Lands	1.000					
X_1	Slope	0.991***	1.000				
X_2	Ruggedness Number	0.992***	0.992***	1.000			
X_3	Mean Annual Rainfall	0.352*	0.349*	0.412*	1.000		
X_4	Drainage Density	0.968***	0.962***	0.970***	0.468*	1.000	
X_5	Quality of Water in *p*H	—0.001*	0.018	—0.012	—0.312	—0.177	1.000
X_6	Behaviour of Water Table	—0.987***	—0.970***	—0.975***	—0.354*	—0.952***	—0.062
X_7	Nitrogen of the Soil	—0.823***	—0.792***	—0.794***	—0.260	—0.757***	—0.359*
X_8	Phosphate of the Soil	—0.969***	—0.951***	—0.970***	—0.378*	—0.972***	0.141
X_9	Potash of the Soil	0.968***	0.967***	0.966***	0.470*	0.987***	—0.145
X_{10}	*p*H Value of the Soil	—0.218	—0.264	—0.267	—0.172	—0.264	—0.019
X_{11}	Distance of the Canal	—0.977***	—0.975***	—0.966***	—0.399*	—0.949***	—0.085

Sl. No.	Variables	X_6	X_7	X_8	X_9	X_{10}	X_{11}
Y	% of Waterlogged Lands						
X_1	Slope						
X_2	Ruggedness Number						
X_3	Mean Annual Rainfall						
X_4	Drainage Density						
X_5	Quality of Water in pH						
X_6	Behaviour of Water Table	1.000					
X_7	Nitrogen of the Soil	0.876***	1.000				
X_8	Phosphate of the Soil	0.961***	0.761***	1.000			
X_9	Potash of the Soil	—0.949***	—0.774***	—0.928***	1.000		
X_{10}	pH Value of the Soil	0.229	0.285	0.227	—0.250	1.000	
X_{11}	Distance of the Canal	0.975***	0.865***	0.916***	—0.968***	0.229	1.000

* Significant at 10% level of significance

** Significant at 5% level of significance

*** Significant at 1% level of significance

3.2.2 Climate (Mean Annual Rainfall X_3)

This indicator is chosen to get an idea about the amount of water that is available for the logging condition though, significantly, it may not be the single contributing agent as excessive irrigational water too can trigger off logging. (Since the data on Mean Annual Rainfall at the village is not recorded, for the sample villages, mean annual rainfall data is generalised at best using the block and tahsil level information). Further there is little variation of rainfall from village to village as given in Table 3.1. Even the result of the correlation matrix (Table 3.2) proves that there is positive correlation between mean annual rainfall, X_3 and waterlogged land Y which is significant at 10% level of significance. Thus the variable mean annual rainfall is not so important in explaining the variation in waterlogged land.

3.2.3 Hydrologic (Drainage Density X_4, Quality of Water in *p*H X_5 and Behaviour of Water Table X_6)

Drainage density length must provide a possible trend with waterlogging conditions. The surfacial drainage lines include the natural and artificial lines, both of which are considered here for the analysis. Drainage density is very high (8.1 to 8.7) in Malapur, Jagdishpur, Asrawan, Katawana villages and very low (1.3 to 3.2) in Mustafabad Saraiya, Kadipur Khurd and Haripur as given in the Table 3.1, where percentage of waterlogged land is very high and very low respectively. In most of the villages where waterlogging is observed to be more, artificial drainage provides the conditions for waterlogging through a rising trend of water table. Thus high drainage density leads to high percentage of waterlogging through a rising trend of water table by both the sub-surface and surface drainage. Table 3.2 proves the above discussed assumption that there is a positive significant relationship between drainage density and percentage waterlogged land which is significant at 1% level of significance.

The irrigation water may also contain salts, and such water, if used frequently, results in the accumulation of salts. Through the seepage the brackish water percolates underground but at a certain level due to hardness, percolation of water gets stuck and continuous percolation of water provides a situation for the rising trend in the water table, due to which waterlogging problem increases. Such type of waterlogging is very scant and is recorded only in usar areas, mainly in Amethi Tahsil. But most of the villages where waterlogging is found to be very high, the quality of water in *p*H (Table 3.1) does not show the same trend because the waterlogging is recorded due to accumulation of water in low-lying areas where it has no outlet. That is why the independent variable, quality of water in *p*H (X_5) has not been considered for further analysis. Even the result of the correlation matrix (Table 3.2) provides that the relationship between waterlogged land (Y) and quality of water in *p*H (X_5) is not significant.

It has been recorded from the Table 3.1 that the water table is found very near to the earth surface (1.3 to 1.5 metres) in Malapur Jagdishpur, Asrawan, Katawana villages and the water table is very low (5.2 to 5.8 metres) from the surface in Musatafabad Saraiya, Kadipur Khurd and Haripur villages where percentage of waterlogged land is very high and very low respectively. Thus there is negative correlation between waterlogged land and behaviour of water table because the values of the water table are correlated in absolute terms. It is proved from the result of the correlation matrix (Table 3.2) that there is negative correlation between waterlogged land and behaviour of water table which is significant at 1% level of significance.

3.2.4 Pedogenic (Nitrogen X_7, Phosphate X_8, Potash X_9 and *p*H Value of the Soil X_{10})

The variables nitrogen, phosphate and potash of the soil in the affected areas of the sample villages are given in the Table 3.1 and they do not provide much variation from village to village and that is why these variables are not considered for further analysis to explain the variations in waterlogged land through stepwise regression analysis. Anyhow these variables are tested through the correlation matrix as given in the Table 3.2.

The *p*H value of the soil is the reading of saturated soil paste, saturation extract, 1 : 1 or 1 : 5 soil water suspension measured with glass and expressed in terms of *p*H scale from 0 to 14. Excessive irrigation water may also contain salts and such water if used frequently results in the accumulation of salts in the soil, which may range from 0.04 to as much as 2 tons of the salts per hectare foot of the water. Owing to use of brackish water through irrigation large areas of the district have been rendered unfit for cultivation. The water table rises nearer to the surface within the range of 1.5 to 2 metres, due to the brackish water which collects in the swamps of low-lying areas. In a few villages accumulation of salt is recorded in waterlogged areas due to which *p*H value of the soil has increased. Thus these two variables may have a positive relationship. But in most villages of the district and even in the sample villages *p*H value of the soil is not high because the accumulated water in the low-laying areas does not contain many salts due to which *p*H value of the soil is found to be low. It is recorded from Table 3.1 that very high *p*H value of the soil (7.73 to 7.79) is found in Haripur, Baijipur and Mebupur Perchhauli villages while very low *p*H value of the soil (7.45 to 7.48) is found in Kamangirpur, Kurapur and Mustafabad Saraiya. The result of the correlation matrix which is depicted in Table 3.2 proves that there is negative correlation between waterlogged land (Y) and *p*H value of the soil (X_{10}) but this relationship is insignificant. So in this case the above discussed hypothesis is not proved.

3.2.5 Anthropogenic (Distance of Canal from Field X_{11})

The villages with very high percentage of waterlogged land of the district lie nearer to the canal and most of these villages are joined through small canal drains for irrigational facilities. Village Malapur Jagdishpur, Kadipur Tahsil, has the affected waterlogged area much nearer to the canal as the canal passes through the village. The affected waterlogged area is 46 metres away from the main canal but it is joined with several small unlined canal drains. The excess water through small drains from the canal and overirrigated water from the field accumulates easily due to nearness of the low-lying area and also due to favourable topographical conditions. It is also recorded that the villages which have low percentage of waterlogged land are too faraway from the canal as given in Table 3.1. Thus there may be negative correlation between percentage of waterlogged land and distance of the canal from the field because it provides better conditions for a rising trend of the water table through seepage and more accumulation of water through both surface and subsurface drainage while outflow does not increase correspondingly. The result of the correlation matrix which is given in Table 3.2 upholds the hypothesis that the distance of the canal from field X_{11} and percentage of waterlogged land Y have negative significant correlation which is significant at 1% level of significance.

3.2.6 Stepwise Regression Analysis

The abovementioned set of environmental and anthropogenic factors have been tested to explain the variation of waterlogged land, which is depicted in Table 3.3, in three sets of combinations of variables, through stepwise regression analysis. Among all the variables 8 variables are significant at 1% level of significance and one variable is significant at 10% level of significance. For the stepwise regression analysis it was not possible to programme all the variables. So with different combinations of variables, stepwise regression analysis is carried out, selecting only those variables which have a significant relationship with the distribution of waterlogged land and which have less multicollinearity with other independent variables.

In the first combination of stepwise regression analysis, variable slope (X_1) enters, in which R^2 is found to be 0.991. In this step, 99.1% variation in waterlogged land is explained while in the second step with the variable X_1, behaviour of water table (X_6) is incorporated and these two variables jointly explain 99.6% variation of waterlogged land which is significant at 1% level of significance. In the third step variabie X_4 (drainage density) is incorporated and jointly with variables X_1 and X_6 explains 99.7% variation of waterlogged land. The 't' value is significant at 1% level of significance and F is also significant at 1% level of significance. In the fourth step variable X_{11} (distance of canal from field) enters which explains 99.7% variation in waterlogged land. It has been recorded that

TABLE 3.3 : **Waterlogged Land : Results of the Stepwise Regression Analysis**

Variables	Intercept	Regression Coefficient	S.E. of R.C.	T	R^2	Increase in R^2	$\bar{R}$	$\bar{R}^2$	F
1	2	3	4	5	6	7	8	9	10
Step 1) (1	—9.21572	10.095	0.377	26.783***	0.991	—	0.99106	0.98219	717.318***
Step 2) (1	6.75542	5.842	1.019	5.731***	0.996	0.005	0.99622	0.99245	850 259***
) (6		—2.048	0.476	—4.299***					
Step 3) (1	6.76569	5.235	1.195	4.380***	0.997	0.002	0.99623	0.99247	565.093***
) (6		—1.913	0.497	—3.849***					
) (4		0.291	0.298	0.849***					
Step 4) (1	3.96668	5.509	1.4000	3.935***	0.997	0.000	0.99596	0.99139	392.282***
) (6		—2.048	0.606	—3.377***					
) (4		0.297	0.310	0.957					
) (11		0.021	0.049	0.425					
Step 1) (1	—0.07914	0.133	0.005	22.246***	0.992	—	0.99249	0.98504	855.316***
Step 2) (1	—0.30771	0.129	0.004	31.265***	0.995	0.003	0.99456	0.98915	589.660***
) (4		0.001	0.000	2.416**					
Step 3) (1	—0.26256	0.129	0.004	29.734***	0.995	0.000	0.99414	0.98831	362.136***
) (4		0.001	0.001	2.102*					
) (5		—0.005	0.021	—0.232					

Step 1) (4	204.22413	1.328	0.695	1.911*	0.468	—	0.46834	0.21934	3.653*
Step 2) (4	173.65584	3.978	2.225	1.788*	0.556	0.088	0.50624	0.25264	2.688*
) (6		2.123	3.296	1.251					
Step 3) (4	183.89882	3.625	2.418	1.500	0.569	0.013	0.45875	0.21045	1.781
) (6		5.925	5.103	1.161					
) (11		—1.179	0.378	—9.474					
Step 4) (4	198.73644	3.378	2.565	1.395	0.569	0.000	0.38401	0.13988	1.199
) (6		5.906	5.351	1.104					
) (11		—0.180	0.396	—0.456					
) (10		—1.893	15.952	—0.119					
Step 5) (4	195.17103	3.429	3.276	1.047	0.570	0.001	0.23358	0.05455	0.865
) (6		6.155	6.431	0.957					
) (11		—0.175	0.422	—0.415					
) (10		—1.697	16.985	—0.100					
) (2		8.682	107.734	0.081					

* Significant at 10% level of significance.
** Significant at 5% level of significance.
*** Significant at 1% level of significance.

at 4th step R^2 is constant and at this step explaining power of the independent variable to dependent variable is knocked down. Thus in the first combination of stepwise regression analysis variable (X_1) slope is an important variable followed by the behaviour of water table X_6, drainage density (X_4) and distance of canal from field (X_{11}), jointly all these variables explain 99.7% of the total variation in waterlogged land.

In the second combination of stepwise regression analysis in the first step, variable X_4 (drainage density) is entered which explains 46.8% variation in waterlogged land significant at 1% level of significance. In the second step variable behaviour of water table (X_6) is incorporated which explains 55.6% variation of waterlogged land, while in the third step variable X_{11} (distance of canal from field) jointly with X_4 and X_6 explains 56.9% variation of waterlogged land.

The third combination of stepwise regression analysis provides that in the first step variable X_1 (slope) is entered where R^2 is 0.992 which explains 99.2% total variation of waterlogged land. In the second step variable drainage density (X_4) is incorporated which jointly with variable X_1 explains 99.5% variation of waterlogged land while in the third step variable X_5 (quality of water in *p*H) jointly with variable X_1 and X_4 explains only 99.5% variation in waterlogged land because at this step R^2 is declined. So the main variables in this combination of stepwise regression are slope and mean annual rainfall but again mean annual rainfall is not so important because at this step increase in R^2 is very nominal which is only 0.003, which alone explains 0.3% variation of waterlogged land.

On the whole 99.6% of the total variation in waterlogged land is explained by the combination of environmental and anthropogenic factors. However the set of environmental and anthropogenic factors and that of waterlogged land provide a vital clue of great significance. The main variables explaining the variation of waterlogged land which is recorded from the above discussion are slope (X_1), behaviour of water table (X_6), drainage density (X_4) and distance of the canal from field (X_{11}) significant at 1% level of significance.

3.3 FACTORS IN USAR LAND

The areas where precipitation is usually less than evaporation in association with elements of topography leading to impeded drainage or with ground water hydrology characterised by high water table enriched in soluble salts in some cases are often responsible for the accumulation of salts either in-situ or of transported nature, thereby causing manifestation of saline and alkali characters in the soil. The prolonged use of saline irrigation water under such conditions also brings about development of soil salinity. The effect of canal irrigation varies according to local condition of the soil, topography, climate and water use. However it has been observed that though the quality of water is generally good the canal irrigation leads to soil salinity in many areas because the canal crosses through many usar areas and it affects the canal water. This problem is mainly due to overirrigation, mismanagement of irrigation water, seepage from canals and rise in water table. The available information shows that many areas which were fertile and productive before canal irrigation are now affected by waterlogging and soil salinity.

The saline soils are found to be distributed throughout India. These are interzonal soils which are characterised by high salinity and sodiumization or both together. It has been estimated that an approximate area of about 7 million hectares is covered by such soils all over the country (Report of Waste Land Survey Committee, Planning Commission, New Delhi). The usar problem has been studied in the past by a number of important Commissions, Government institutions and various soil scientists. This has been discussed in the literature survey of the first chapter, where it is pointed out that the probe into the causative factors for the development of these lands has been dealt with scant attention.

The 15 villages which are selected on the quartiling basis, to analyse responsible factors in the distribution of usar land, in which very high percentage usar land villages (33.96 to 39.96 to the total area of the village which is depicted in Appendix 2) are Ghazipur, Purainachandrabhan, Dilipgarh; high percentage (21.43 to 32.18) usar land villages are Bhainsada, Bhojpur, Kashipur; medium percentage (12.21 to 17.85) villages are

Malabadospati, Tewaripur and Bhain. Low percentage (6.16 to 9.88) usar land sample villages are Purkhipur, Sherpur and Nonara and very low percentage (0.74 to 3.00) of usar land sample villages are Kharsoma, Mirdadpur and Baithu.

As in the waterlogged land a set of environmental and anthropogenic parameters are selected to find out their impact on these lands with the help of correlation matrix and regression analysis. For this purpose percentage distribution of usar land, in the sample villages as given above, is taken as dependent variable and morphometric, climatic, hydrologic and pedogenic parameters of environment are taken as independent variables.

3.3.1 Morphometric (Slope X_1 and Ruggedness Number X_2)

The areas with greater heights do account for greater slope and greater ruggedness number. Hence both these variables are correlated with each other. It may be seen from Table 3.4 that the distribution pattern of slope (X_1) and ruggedness number (X_2) is the same in the selected sample usar land villages. It is also seen that the villages which have a high percentage of usar land, i.e. Ghazipur, Purainachandrabhan and Dilipgarh, have a very low slope (1.78° to 1.85°) and ruggedness number while villages having a very low percentage of usar land have a very high slope and ruggedness number which means that there is a negative relationship between dependent variable-percentage of usar land and independent variables-slope and ruggedness number because the area with a low slope and ruggedness number helps in accumulation of natural rainy water or salty irrigation water and in the dry season presence of salt is found on the surface. The sub-soil water in the affected area is charged with heavy concentration of salts and with the rise of sub-soil water by capillary action, these are brought up to the earth surface which is a hindrance to the crop production. The assumption of slope (X_1) and ruggedness number (X_2) prove (Tabel 3.5) that there is significant negative correlation between usar land (Y dependent variable) and slope (X_1) and ruggedness number (X_2) (independent variables) which are significant at 1% level of significance.

TABLE 3.4 : **Factors in usar land, 1981-82**

Sample villages	% of usar land to total area	Slope in degrees	Ruggedness number	Mean Annual Rainfall in mms.	Drainage Density	Quality of water in pH	Behaviour of water table in feet	Pedogenic: Nitrogen	Pedogenic: Phosphate	Pedogenic: Potash	Pedogenic: pH Value	Thickness of the top soil in metres
	(Y)	(X_1)	(X_2)	(X_3)	(X_4)	(X_5)	(X_6)	(X_7)	(X_8)	(X_9)	(X_{10})	(X_{11})
Ghazipur	39.96	1.78	0.18	175	2.27	9.82	10	2.17	1.69	2.99	10.50	1.24
Purainachandrabhan	37.89	1.82	0.18	216	2.30	9.73	12	2.15	1.67	2.95	10.35	1.35
Dilipgarh	33.33	1.85	0.18	177	2.31	9.65	15	2.11	1.64	2.94	10.26	1.52
Bhainsada	32.18	1.89	0.19	175	2.36	9.77	17	2.09	1.62	2.93	10.10	1.71
Bhojpur	26.93	1.92	0.19	172	2.42	9.29	20	2.05	1.60	2.90	9.95	1.83
Kashipur	21.43	1.95	0.19	188	2.47	9.08	21	2.02	1.58	2.87	9.85	1.94
Malakdoshipatti	17.85	1.98	0.20	212	2.50	8.95	23	1.99	1.54	2.83	9.75	2.04
Tewaripur	14.29	2.01	0.20	185	2.57	8.76	25	1.95	1.51	2.79	9.67	2.08
Bhain	12.21	2.04	0.20	168	2.68	8.47	27	1.94	1.48	2.71	9.52	2.10
Purkhipur	9.88	2.07	0.20	208	2.92	8.38	30	1.86	1.42	2.68	9.43	2.14
Sherpur	7.14	2.10	0.21	189	3.10	8.15	32	1.82	1.38	2.64	9.21	2.17
Nonara	6.16	2.15	0.21	178	3.25	7.96	33	1.75	1.35	2.61	9.10	2.18
Kharsoma	3.00	2.22	0.22	190	3.43	7.82	35	1.71	1.30	2.52	8.81	2.19
Mirdadpur	1.57	2.27	0.22	194	2.54	7.73	36	1.67	1.26	2.45	8.52	2.20
Baithu	0.74	2.32	0.23	190	4.23	7.61	38	1.62	1.21	2.38	8.30	2.22

TABLE 3.5 : **Correlation Matrix**

Sl. No.	Variables	Y	X_1	X_2	X_3	X_4	X_5
Y	% of the Usar Lands	1.000					
X_1	Slope	—0.962***	1.000				
X_2	Ruggedness Number	—0.942***	0.984***	1.000			
X_3	Mean Annual Rainfall	—0.161	0.175	0.178	1.000		
X_4	Drainage Density	—0.860***	0.957***	0.955***	0.196	1.000	
X_5	Quality of Water in *p*H	0.984***	—0.976***	—0.954***	—0.165	—0.914***	1.000
X_6	Behaviour of Water Table	—0.991***	0.982***	0.962***	0.149	0.905***	—0.987***
X_7	Nitrogen of the Soil	0.965***	—0.995***	—0.977***	—0.158	—0.957***	0.986***
X_8	Phosphate of the Soil	0.958***	—0.994***	—0.979***	—0.183	—0.964***	0.981***
X_9	Potash of the Soil	0.942***	—0.991***	—0.974***	—0.192	—0.972***	0.973***
X_{50}	*p*H Value of the Soil	0.949***	—0.998***	—0.985***	—0.178	—0.965***	0.969***
X_{11}	Thickness of the Top Soil	—0.994***	0.969***	0.851***	0.108	0.707***	—0.893***

S. No.	Variables	X_6	X_7	X_8	X_9	X_{10}	X_{11}
Y	% of the Usar Lands						
X_1	Slope						
X_2	Ruggedness Number						
X_8	Mean Annual Rainfall						
X_4	Drainage Density						
X_5	Quality of Water in *p*H						
X_6	Behaviour of Water Table	1.000					
X_7	Nitrogen of the Soil	—0.985***	1.000				
X_8	Phosphate of the Soil	—0.981***	0.998***	1.000			
X_9	Potash of the Soil	—0.966***	0.992***	0.995***	1.000		
$\bar{X}_{19}$	*p*H Value of the Soil	—0.973***	0.992***	0.992***	0.992***	1.000	
X_{11}	Thickness of the Top Soil	0.931***	—0.863***	—0.848***	—0.818***	—0.850***	1.000

* Significant at 10% level of significance.
** Significant at 5% level of significance.
*** Significant at 1% level of significance.

3.3.2 Climatic (Mean Annual Rainfall X_3)

As it has been emphasized earlier that the data for the mean annual rainfall is collected at the tahsil and block level and the available information is generalised for the selected sample villages. Even the variation of mean annual rainfall at village and block level is not so much that it can be considered for further analysis which is given in Table 3.4 and it has been recorded from Table 3.5 that the relationship between usar land and mean annual rainfall is not significant. But the climate in general has an utmost impact on usar land as Sultanpur District falls in a semi-arid tract.

3.3.3 Hydrologic (Drainage Density X_4, Quality of Water in *p*H X_5 and Behaviour of Water Table X_7)

It has been recorded from Table 3.4 that the villages which have a very high percentage of usar land have a very low drainage density, while drainage density is very high in these villages where the percentage of usar land is very low, because along with less amount of drainage lines, the evaporation also exceeds and the salts get encrusted on the surface. But, on the other hand, if there are more drainage lines, there is every possibility of salts being washed away with the flowing water. It is proved from the correlation matrix (Table 3.5) that there is a negative significant correlation between percentage of usar land Y and drainage density X_4 which is significant at 1% level of significance.

It has been observed from Table 3.4 that the villages having a high percentage of usar land have a very high quality of water in *p*H and when the percentage of usar land is low, quality of water in *p*H is also low, due to the collection of brackish water with high salts. The brackish water dries up in two ways, firstly, through evaporation and, secondly, through percolation. It is clear from the second process that when the salty water percolates downward it is distributed in the underground water. On the other hand when the poor quality of water, i.e. the same salty underground water, is used for irrigation purposes, salt again spreads over the land, through which usarisation is recorded in the field. The above assumption is proved from the correlation matrix (Table 3.5) that there is a positive significant correlation between quality of water in *p*H (X_5) and percentage of usar land (Y) which is significant at 1% level of significance.

The behaviour of water table is observed to be very high in those villages where the percentage of usar land is also very high, and where the percentage of usar land in the sample villages is low, the water table is found much below because at a certain strata the solid profile is little permeable due to restrictions of kankar pan or indurated clay pan in the sub-soil, and due to this water table rises which is enriched in soluble salts. This is responsible for the accumulation of salts either in-situ or transported and deposited in other places thereby causing usarisation in the soil. It is also statistically proved from the correlation matrix (Table 5) that there is significant negative correlation at 1% level of significance between percentage of usar land Y and behaviour of water table X_6.

3.3.4 Pedogenic (Nitrogen X_7, Phosphate X_8, Potash X_9, *p*H Value of the Soil X_{10} and Thickness of the top Soil X_{11})

The variables nitrogen, phosphate and potash of the soil have mainly been taken to estimate the fertilizer requirement of the soil, and firstly an attempt is made to test these variables at block level with the total waste lands, which provides significant relationship because the less variation in nitrogen, phosphate and potash of the soil either shows an increasing or decreasing trend among the sample villages but these variables are not considered for the further analysis, to explain the variation in usar land.

It has been explained earlier that the *p*H value of the soil is the reading of the saturated soil paste, saturation extract 1 : 1 or 1 : 5 soil water suspension, measured with glass electrode and expressed in terms of *p*H scale from 0 to 14. Thus in the usar soils, due to excessive concentration of neutral salts mainly of chlorides and sulphates of sodium, *p*H value of the soil is high and this is clear from Table 3.4 that a very high percentage of sample villages have a very high *p*H value of the soil and it is also proved from the correlation matrix (Table 3.5) that there is positive significant correlation between usar land and *p*H value of the soil which is significant at 1% level of significance.

It is observed from Table 3.4, that the villages with a very high percentage of usar land have very low thickness of the top soil, because in soils with less thickness of top soil, leaching and percolation of water is possible and with this rise in water table is also recorded high, which provides suitable conditions for the usarisation. The result of the correlation matrix (Table 3.5, provides clues that there is a negative significant correlation between percentage of usar land (Y) and thickness of top soil (X_{11}) which is significant at 1% level of significance.

3.3.5 Stepwise Regression Analysis

The system of explanation by X (independent variable) discussed above shows a high degree of multi-collinearity among themselves. To remove the inconsistencies arising out of this problem of multi-collinearity a stepwise regression analysis is attempted.

In this exercise percentage of usar land is considered as dependent variable (Y) and other environmental variables mentioned earlier as independent variables (X). The result of the stepwise regression analysis gives an inter-correlation matrix which is given in Table 3.6 in three sets of stepwise regression. In various sets of stepwise regression, after a certain step R^2 starts declining, therefore the analysis is truncated at those steps where R^2 starts declining. Before R^2 starts declining the entered variables are found optimal on the regression line.

The most dominant variables in the first set of stepwise regression in explaining the variability in usar land is the behaviour of water table (X_6) which explains 98.2% variation of the usar land. It is interesting to note that in subsequent steps variable (X_4) drainage density enters and jointly with variable (X_6) explains 99.5% variation of usar land. But in the third step variable (X_3) enters where R^2 is found stagnant instead of increasing which shows that mean annual rainfall has no explaining power of the usar land variation among the sample villages.

TABLE 3.6: Usar Land : Results of the Stepwise Regression Analysis

Variables	Intercept	Regression Coefficient	S.E.	T	R^2	Increase in R^2	$\bar{R}$	$\bar{R}^2$	F
1	2	3	4	5	6	7	8	9	10
Step 1) (6	1.57777	0.018	0.001	18.618***	0.982	—	0.98176	0.96385	346.631***
Step 2) (6	1.44355	0.012	0.001	9.420***	0.995	0.003	0.99464	0.98931	528.372***
) (4		0.102	0.018	5.630***					
Step 3) (6	1.43323	0.012	0.001	9.023***	0.995	0.000	0.99421	0.98845	366.741***
) (4		0.102	C.019	5.314***					
) (3		0.000	0.000	0.178					
Step 1) (10	0.42112	—0.023	0.001	—20.478***	0.985	—	0.98485	0.96993	419.351***
Step 2) (10	0.40045	—0.022	0.002	—10.031***	0.985	0.000	0.98408	0.96841	198.623***
) (11		0.002	0.004	0.553***					
Step 3) (10	0.40418	—0.024	0.005	—4.820***	0.985	0.000	0.98297	0.96623	122.977***
) (11		0.004	0.006	0.650					
) (5		0.002	0.005	0.375					
Step 4) (10	0.40267	—0.024	0.005	—4.582***	0.985	0.000	0.98142	0.96318	33.930***
) (11		0.004	0.006	0.626					
) (5		0.002	0.005	0.359					
) (3		0.000	0.000	0.097					

TABLE 3.6 (*Contd.*)

1	2	3	4	5	6	7	8	9	10
Step 1) (10	4.38871	—0.247	0.004	—58.798***	0.998	—	0.99813	0.99626	3457.217***
Setp 2) (10	4.16120	—0.231	0.006	—36.887	0.999	0.001	0.99885	0.99770	2805.317***
) (11		0.039	0.013	3.011***					
Step 3) (10	3.61813	—0.190	0.028	—6.801***	0.999	0.000	0.99897	0.99794	2073.153***
) (11		0.065	0.021	3.062***					
) (4		0.034	0.023	1.516					
Step 4) (10	3.61915	—0.190	0.029	—6.484***	0.999	0.000	0.99888	0.99776	1413.791***
) (11		0.065	0.022	2.919**					
) (4		0.034	0.024	1.446					
) (3		—0.000	0.000	—0.044					

* Significant at 10% level of significance.
** Significant at 5% level of significance.
*** Significant at 1% level of significance.

In the second set of stepwise regression analysis, the most dominant variable in explaining variation of usar land is *p*H value of the soil (X_{10}) which alone explains 98.5% variation of usar land, while in the second step variable (X_{11}) thickness of the top soil enters and jointly with variable (X_{10}) explains 98.5% variation of usar land because at this step R^2 is constant, and in the third step variable (X_5) quality of water in *p*H and variable (X_3) mean annual rainfall enter and in these steps R^2 is recorded constant, which provides the clue that (X_{11}), (X_5) and (X_3) variables have no explanatory power of the variation of usar land. But the main variable in this set is *p*H value of the soil (X_{10}) which explains the 98.5% variation of usar land.

The most dominant variable, in the third set of stepwise regression analysis, in explaining variation of usar land, is *p*H value of the soil (X_{10}) which explains the 99.8% variation of usar land, while in the second step variable (X_{11}) thickness of the top soil enters and explains the 99.9% variation of usar land jointly with variable (X_{10}). But in the third step variable (X_4) drainage density and variable (X_3) mean annual rainfall in the fourth step enter but in these steps the explaining power of these variables is knocked down because R^2 is found to be constant and it explains the same percentage variation of usar land as it was explained in the second step of stepwise regression. So the main variables in this set are *p*H value of the soil and thickness of the top soil.

Thus the main subsequent variables in all these three sets of stepwise regression analysis are variable X_{10} (*p*H value of the soil), X_6 (behaviour of water table), X_4 (drainage density) and variable X_{11} (thickness of the top soil). The final form of multiple regression analysis is given in three sets of stepwise regression (Table 3.6) where abovementioned variables explain 99.9% of total variation of usar land where F value is found to be significant at 1% level of significance.

3.4 FACTORS IN BANJAR LAND

The information gathered from the revenue officials of the district provides that banjar land is such type of land which is left uncultivated for more than five years and it is also observed from the field that leaving such type of land uncultivated is mainly due to low fertility status of the soil which is unfit to produce crops.

The 15 villages which are chosen for the analysis of the factors on the quartiling basis are: Very High percentage (34.04 to 38.34) villages are—Piparpur, Gulalpur and Kushela; High percentage (29.01 to 32.69) villages are—Magrahi, Gharwaspur and Baksara; Medium percentage (20 to 24.35) villages are—Maj, Dhabawa and Palpur; Low percentage (8.77 to 15.05) villages are—Singhauli, Banbhukar and Bhagirath-patti; and Very Low percentage (1.77 to 5.39) villages are—Baini, Chaukia and Ghoraha.

The abovementioned sample villages are selected for analysing the distribution of banjar land (Appendix 3). The main factors which are responsible for the variation of banjar land and considered for the analysis are morphometric, climatic, hydrologic, pedogenic and anthropogenic.

3.4.1 Morphometric (Slope X_1 and Ruggedness Number X_2)

The variable slope (X_1) and ruggedness number (X_2) show a similar pattern of distribution in the sample villages as given in Table 3.7 and these are interrelated because the areas with high altitudes do account for both greater amount of slope and ruggedness number. It is recorded that the villages having a high percentage of banjar land also have a high amount of slope and ruggedness number, while the variation of these independent variables in sample villages is not high. This is due to the fact that a high slope and ruggedness number provide favourable conditions for erosion due to which top soil fertility is washed away. It has been observed in the field that there is unevenness within the affected area, due to which applied fertilizer and irrigated water gets accumulated in lowlying areas, and due to poor bunding fertility of the soil is washed away to other areas. The above analysis proves statistically (Table 3.8) that there is significant positive correlation-between dependent variable Y (banjar land) and independent variables (slope X_1 and ruggedness number X_2) which are significant at 1% level of significance.

TABLE 3.7 : Factors in Banjar land, 1981-82

Sample Villages	% of Banjar land to total area	Slope in degrees	Ruggedness number	Mean annual rainfall in mms.	Drainage density	Quality of water in pH	Behaviour of water table in feet	Soil: Nitrogen	Soil: Phosphate	Soil: Potash	Soil: pH Value	Land concentration/ Gini's coefficient
	(Y)	(X_1)	(X_2)	(X_3)	(X_4)	(X_5)	(X_6)	(X_7)	(X_8)	(X_9)	(X_{10})	(X_{11})
Piparpur	38.34	1.88	0.28	190	4.78	8.75	15	1.90	1.58	3.14	8.89	0.78
Gulalpur	38.10	1.82	0.28	210	4.37	8.67	13	1.91	1.59	3.10	8.87	0.75
Kushela	34.04	1.70	0.27	215	4.64	8.58	13	1.93	1.60	3.05	8.75	0.71
Magrahi	32.69	1.62	0.27	217	4.50	8.42	13	1.95	1.68	3.00	8.68	0.68
Gharwaspur	32.01	1.51	0.26	205	4.34	8.45	13	1.98	1.69	2.92	8.65	0.67
Baksara	29.01	1.43	0.25	225	4.24	8.56	10	2.00	1.71	2.90	8.52	0.65
Palpur	24.35	1.36	0.24	165	4.11	8.58	20	2.08	1.75	2.86	8.47	0.61
Dhabawa	22.02	1.27	0.24	198	3.98	8.49	15	2.17	1.81	2.78	8.33	0.58
Majhui	20.00	1.19	0.23	207	3.57	9.35	12	2.20	1.84	2.70	8.20	0.53
Singhauli	16.05	1.07	0.22	205	3.52	8.23	10	2.26	1.89	2.63	8.18	0.50
Banbhukar	14.83	0.94	0.20	212	3.46	8.04	10	2.31	1.92	2.54	8.02	0.47
Bhagirathpatti	8.70	0.87	0.19	218	3.31	7.97	10	2.37	1.97	2.47	7.92	0.45
Baini	5.39	0.71	0.18	220	3.20	7.83	10	2.42	2.01	2.38	7.87	0.41
Chaukia	3.50	0.63	0.17	200	3.08	7.85	10	2.56	2.08	2.21	7.75	0.38
Ghoraha	1.77	0.54	0.16	190	2.67	7.79	14	2.68	2.15	2.13	7.67	0.32

3.4.2 Climatic (Mean Annual Rainfall X_3)

There is very little variation in the mean annual rainfall from one village to another, which is given in Table 3.7, for which the data is generalised from the tahsil and block level information. That is why variable mean annual rainfall has not been considered for further analysis and the result of the correlation matrix (Table 3.8) provides enough clue that the relationship between percentage of banjar land and mean annual rainfall is insignificant.

3.4.3 Hydrologic (Drainage Density X_4, Qaulity of Water in *p*H X_5 and Behaviour of Water Table X_6)

The sample villages which have a high percentage of banjar land also have high drainage density (Table 3.7) and low percentage sample villages have low drainage density because it is observed in the field that there is much overland flow in the form of drains either natural or artificial due to which fertility of the soil is being eroded by surface drainage and farmers are bound to leave the land fallow or uncultivated because of their socio-economic limitations. The positive relationship of these two variables is proved from (Table 3.8), that the relationship between banjar land and drainage density is significant at 1% level of significance.

The quality of water in *p*H is recorded very high (Table 3.7) in those villages where percentage of banjar land is very high and villages having a very low percentage of banjar land have very low quality of water in *p*H. This shows that there is positive correlation between percentage distribution of banjar land (Y) and quality of water in *p*H (X_5) because due to the poor quality of ground water when it is used for irrigational purpose, the salt is distributed in the field and after some time fertility of the soil gets deteriorated due to accumulated salts and land is left uncultivated. The above assumption is statistically proved (Table 3.8), that there is significant positive correlation between banjar land and quality of water in *p*H, which is significant at 1% level of significance.

It has been observed (Table 3.7) that the sample villages having a high percentage of banjar land have a low water

TABLE 3.8 : **Correlation Matrix**

Sl. No.	Variables	Y	X_1	X_2	X_3	X_4	X_5
1	2	3	4	5	6	7	8
Y	% of the Banjar_Land	1.000					
X_1	Slope	0.994***	1.000				
X_2	Ruggedness Number	0.992***	0.993***	1.000			
X_3	Mean Annual Rainfall	0.19 ***	—0.002	0 021	1.000		
X_4	Drainage Density	0.988***	0.991***	0.984***	0.034	1.000	
X_5	Quality of Water	0.950***	0.954***	0.958***	—0.169	0.938***	1.000
X_6	Behaviour of Water Table	0.418*	0.437*	0.424*	—0.808***	0.415*	0.547*
X_7	Nitrogen of the Soil	—0.981***	—0.978***	—0.985***	—0.082	—0.981***	—0.980***
X_8	Phosphate of the Soil	—0.991***	—0.993***	—0.989***	—0.038	—0.992***	—0.953***
X_9	Potash of the Soil	0.987***	0.991***	0.993***	0.23	0.985***	0.960***
X_{10}	*p*H value of the Soil	0.995***	0.996***	0.991***	—0.011	0.993***	0.949***
X_{11}	Land Concentration	0.992***	0.996***	0.988***	0.012	0.996***	0.952***

TABLE 3.8 (*Contd.*)

Sl. No.	Variables	X_6	X_7	X_8	X_9	X_{10}	X_{11}
		9	10	11	12	13	14
Y	% of the Banjar Lands						
X_1	Slope						
X_2	Ruggedness Number						
X_3	Mean Annual Rainfall						
X_4	Drainage Density						
X_5	Quality of Water						
X_6	Behaviour of Water Table	1.000					
X_7	Nitrogen of the Soil	—0.379*	1.000				
X_8	Phosphate of the Soil	—0.410*	0.992***	1.000			
X_9	Potash of the Soil	0.430*	—0.994***	—0.996***	1.000		
X_{10}	pH value of the Soil	0.448*	—0.980***	—0.992***	0.988***	1.000	
X_{11}	Land concentration	0.422*	—0.984***	—0.995***	0.990***	0.996***	1.000

* Significant at 10% level of significance.
** Significant at 5% level of significance.
*** Significant at 1% level of significance.

table, in those villages where the percentage of banjar land is low, the water table is high because in the very low water table areas, it is found that soil moisture is very low through the percolation (leaching), due to which crop producing capacity of the soil is reduced. The above hypothesis is statistically not supported strongly. However there is positive relationship between the behaviour of water table and percentage of banjar land (Table 3.8) because it is significant at 10% level of significance.

3.4.4 Pedogenic (Nitrogen X_7, Phosphate X_8, Potash X_9 and *p*H Value of the Soil X_{10})

The variables nitrogen, phosphate and potash of the soil are not considered for further analysis as there is little variation in these variables from one village to another. These have been used to know the capacity of the soil to produce the crop and fertilizer requirement of the soil.

The *p*H value of the soil is found very high in those villages where the percentage of banjar land is also high (Table 3.7) which shows that there is positive correlation between percentage distribution of banjar land and *p*H value of the soil because the soils with high pH value have an excessive amount of salts which is a hindrance to the crop production. Table 3.8 provides enough clue that there is positive correlation significant at 1% level of significance between *p*H value of the soil and percentage distribution of banjar land.

3.4.5 Anthropogenic (Land Concentration X_{11} or Gini's Coefficient)

Land concentration is recorded very high (0.71 to 0.78), which is given in (Table 3.7), in those villages where percentage of banjar land is low, land concentration is also low, which shows that there is positive correlation between these variables because the large size land holding farmers are least bothered to improve the fertility status of the soil because, on the one hand, they are getting less output and, on the other, they have other better lands for agricultural use. So they apply much input to the good quality land only where they are sure of getting good profits and that is the reason why the banjar land are left uncultivated. The above assumption is statistically proved (Table 3.8), that there is significant positive correlation, significant at 1% level of significance, between percentage distribution of banjar land and land concentration.

3.4.6 Stepwise Regression Analysis

In the abovementioned correlation matrix 11 parameters of environmental and anthropogenic factors have been tested among which some of them show a positive trend with banjar land and others show a negative trend. For the stepwise regression analysis it was not possible to programme all the variables, so in the various sets, stepwise regression analysis is attempted, selecting those variables which show less multi-collinearity and have significant relationship with the banjar land as depicted in Table 3.9.

In the first set of stepwise regression analysis, in the first step variable (X_{10}) *p*H value of the soil enters which explains 99.5% total variation of the banjar land, while in the second step variable X_1 (slope) jointly with variable X_{10} explains 99.5% variation of the banjar land. After this in the third step variable X_3 (mean annual rainfall), in the fourth step variable X_6 (behaviour of water table) and in the fifth step variable X_{11} (land concentration) are incorporated respectively and these variables have no explanation for banjar land distribution because in these steps R^2 is constant instead cf increasing and in these steps abovementioned variables have less importance and the only main variables are the *p*H value of the soil and slope which explain the variation of the banjar land.

The variable X_{11} (land concentration) is recorded as dominant variable (Table 3.9), in the second set of stepwise regression which alone explains 99.2% variation of the banjar land, while in the second step variable X_2 (slope) jointly with variable X_{11} explains 99.5% variation of the banjar land and in the third step variable X_5 (quality of water in *p*H) is incorporated and, in combination with variable X_{11} and X_2, explains 99.6% total variation of the banjar land. Lastly, in the fourth step variable X_4 (drainage density) and in the fifth step variable X_3 (mean annual rainfall) are incorporated and in these steps explaining

TABLE 3.9 : **Banjar Land : Results of the Stepwise Regression Analysis**

Variables	Intercept	Regression coefficient	S.E.	T	R_2	Increase in R^2	$\bar{R}$	$\bar{R}^2$	F
1	2	3	4	5	6	7	8	9	10
Step 1) (10	—233.57454	30.652	0.895	34.236***	0.995	—	0.99450	0.98903	1172.117***
Step 2) (10	—127.55626	15.813	9.096	1.739*	0.996	0.001	0.99516	0.99034	663.344***
) (1		14.086	8.597	1.638					
Step 3) (10	—135.78922	17.150	9.226	1.859*	0.996	0.000	0.99517	0.99036	440.069***
) (1		13.207	8.669	1.523					
) (6		—0.136	0.141	—0.966					
Step 4) (10	—145.01375	18.654	11.276	1.654	0.996	0.000	0.99476	0.98954	302.050***
) (1		14.669	10.695	1.372					
) (6		—0.147	0.154	—0.959					
) (11		—8.736	33.947	—0.257					
Step 5) (10	—145.86297	18.527	11.926	1.553	0.996	0.000	0.99425	0.98853	217.811***
) (1		14.633	11.269	1.299					
) (6		—0.110	0.356	—0.309					
) (11		—8.577	35.872	—0.240					
) (3		0.007	0.057	0.117					

TABLE 3.9 (*Contd.*)

1	2	3	3	5	6	7	4	9	10
Step 1) (11	— 28.63798	88.383	3.026	29.205***	0.992	—	0.99247	0.98499	852.932***
Step 2) (11	— 39.18866	44.494	15.915	2.796**	0.995	0.003	0.99509	0.99020	653.091***
) (2		154.325	55.294	2.791**					
Step 3) (11	— 26.64802	45.540	16.586	2.746**	0.996	0.001	0.99479	0.98960	407.814***
) (2		165.227	61.360	2.693**					
) (5		—1.883	3.862	—0.488					
Step 4) (11	— 20.67379	55.350	34.932	1.585	0.996	0.000	0.99437	0.98877	280.987***
) (2		169.237	65.211	2.595**					
) (5		—2.419	4.357	—0.555					
) (4	— 2.053	5.350	—0.323						
Step 5) (11	— 13.08490	55.583	36.697	1.515	0.996	0.000	0.99385	0.98774	203.823***
) (2		175.560	72.731	2.414					
) (5		—3.304	5.717	—0.578					
) (4		—2.068	6.669	—0.310					
) (3		—0.009	0.033	—0.258					

* Significant at 10% level of significance.
** Significant at 5% level of significance.
*** Significant at 1% level of significance.

power of the variables is knocked down because the R^2 is constant. Thus the mean variables in this set of stepwise regression analysis are land concentration, slope and quality of water in *p*H.

Thus the total variation in banjar land explained by the environmental and anthropogenic factors is 99.6%. However the interaction between the set of environmental and anthropogenic factors on the one hand and that of the banjar land on the other provide a clue of great significance at 1% level of significance, while the main variables in explaining the percentage distribution of banjar land are land concentration, *p*H value of the soil, slope and quality of water in *p*H.

3.5 Factors in Old Fallow Land

The old fallow land includes that land which is left uncultivated for 2 to 5 years for some reason or the other. Investigations from the Lekhapals and other revenue officials and villagers reveal that this is such type of land as is not taken up for cultivation due to deteriorated condition of the soil. In this analysis a few environmental and anthropogenic factors are considered to analyse the distribution of old fallow land.

For this purpose 15 villages are selected on the basis of quartiling (see Appendix 4) : Very High percentage (41.90 to 44.51) of old fallow land villages are Raibigo, Bhojpur and Bhagupur; High percentage (33.35 to 39.28) villages are Raghaipur, Richaura and Mahmudpur; Medium percentage (17.84 to 28.37) villages are Mujesh, Dharmdashpur and Sakardepur; Low percentage (8.64 to 14.39) villages are Saranghpur; Sirwara and Karondipur. Very Low percentage (1.84 to 5.73) villages are Dhundhu, Gariyanawa and Tirahi.

3.5.1 Morphometric (Slope X_1 and Ruggedness Number X_2)

The distribution pattern of slope and ruggedness number are observed to be the same (Table 3.10) and these two variables are recorded very high in those villages where the percentage of old fallow land is high, and there seems to be a relationship between the two because where the percentage of old fallow land is low, slope and ruggedness number are also recorded

TABLE 3.10 : **Factors in Old Fallow Land, 1981-82.**

Sample Villages	% old fallow land to total area	Slope in degrees	Ruggedness Number	Mean Annual Rainfall in mms.	Drainage Density	Quality of water in *p*H	Behaviour of water table in feet	Soil: Nitrogen	Soil: Phosphate	Soil: Potash	Soil: *p*H Value	Land Concentration' Gini's coefficient
	(Y)	(X_1)	(X_2)	(X_3)	(X_4)	(X_5)	(X_6)	(X_7)	(X_8)	(X_9)	(X_{10})	(X_{11})
Raibigo	44.41	2.12	0.31	187	6.72	7.98	30	1.95	1.69	2.91	7.68	0.67
Bhojpur	43.56	2.12	0.31	195	6.70	7.96	28	1.97	1.70	2.94	7.56	0.65
Bhagupur	41.90	2.10	0.30	155	6.62	7.94	27	1.99	1.72	2.97	7.65	0.62
Raghaipur	39.28	2.07	0.30	190	6.58	7.91	26	2.07	1.75	3.01	7.64	0.60
Richhaura	34.60	2.05	0.29	158	6.47	7.87	25	2.09	1.79	3.07	7.63	0.57
Mahmudpur	32.35	2.02	0.28	186	6.32	7.83	24	2.13	1.82	3.09	7.57	0.52
Mujesh	28.37	1.98	0.27	195	6.26	7.80	23	2.17	1.87	3.13	7.56	0.48
Dharmdashpur	22.73	1.95	6.27	190	6.17	7.76	22	2.19	1.89	3.17	7.54	0.43
Sakardepur	19.84	1.90	0.26	195	6.05	7.75	20	2.23	1.93	3.21	7.52	0.39
Saranghipur	14 33	1.83	0.26	197	5.87	7.72	19	2.35	1.95	3.30	7.50	0.37
Sirwara	10.14	1.76	0.25	190	5.37	7.71	18	2.38	1.97	3.37	7.45	0.32
Karondipur	8.68	1.71	0.24	185	4.86	7.67	17	2.43	2.01	3.39	7.44	0.28
Dhundu	5.73	1.64	0.23	197	4.57	7.65	14	2.48	2.08	3.42	7.43	0.21
Gariynawa	3.06	1.58	0.22	205	4.32	7.52	12	2.56	2.13	3.47	7.41	0.19
Tirahi	1.84	1.46	0.20	195	4.20	7.51	10	2.69	2.17	3.51	7.40	0.11

low. In most of the villages it is observed that with a high slope and ruggedness number, land is liable to suffer from erosion, due to which fertility of the soil goes down and the land is usually left uncultivated. The result of the correlation matrix (Table 3.11) proves that there is a significant positive correlation between the percentage of old fallow land and slope and ruggedness number which are significant at 1% level of significance.

3.5.2 Climatic (Mean Annual Rainfall X_3)

The independent variable mean annual rainfall as not considered for further analysis because there is little variation in the mean annual rainfall at village level (Table 3.10) and data is also based on the tahsil and block level information. Even it is proved from the correlation matrix (Table 3.11) that there is negative correlation between percentage of old fallow land and mean annual rainfall and it has less importance because these two variables are significant at 1% level of significance.

3.5.3 Hydrologic (Drainage Density X_4, Quality of Water in *p*H X_5 and Behaviour Water Table$_6$)

Drainage density is found to be very high (Table 3.10) in those villages where the percentage of old fallow land is very high and it is low in those villages where the percentage of old fallow land is low because high drainage density leads to more erosion of the top soil by which fertility decreases and the farmers are bound to leave the land uncultivated. It is also proved statistically (Table 3.11) that there is a positive significant correlation at 1% level of significance between percentage of old fallow land and drainage density.

Underground brackish (poor quality) water used for irrigation through pumping disturbs good agricultural land, spreading salt on the surface and by this crop producing capacity of the soil is hindered and that is why the quality of water in *p*H is recorded very high (Table 3.10) in those villages where the percentage of old fallow land is very high and it is low in those villages where the percentage of old fallow land is low. It is also peroved from Table 3.11 that there is positive significant

TABLE 3.11 : **Correlation Matrix**

Sl. No.	Variables	Y	X_1	X_2	X_3	X_4	X_5
1	2	3	4	5	6	7	8
Y	% of the Old Fallow Lands	1.000					
X_1	Slope	0.965***	1.000				
X_2	Ruggedness Number	0.971***	0.986***	1.000			
X_3	Mean Annual Rainfall	—0.536*	—0.627**	—0.614**	1.000		
X_4	Drainage Density	0.939***	0.986***	0.964***	—0.579**	1.000	
X_5	Quality of Water in *p*H	0.972***	0.973***	0.988***	—0.628**	0.948***	1.000
X_6	Behaviour of Water Table	0.976***	0.988***	0.992***	—0.618**	0.966***	0.988***
X_7	Nitrogen of the Soil	—0.975***	—0.994***	—0.989***	0.623**	—0.969***	—0.983***
X_8	Phosphate of the Soil	—0.583***	—0.984***	—0.995***	0.629**	—0.962***	—0.993***
X_9	Potash of the Soil	—0.996***	—0.978***	—0.979***	0.570**	—0.959***	—0.977***
X_{10}	*p*H Value of the Soil	0.994***	0.964***	0.975***	—0.601**	0.943***	0.971***
X_{11}	Land Concentration	0.988***	0.987***	0.994***	—0.619**	0.964***	0.985***

Sl. No.	Variables	X_6	X_7	X_8	X_9	X_{10}	X_{11}
		9	10	11	12	13	14
Y	% of the Old Fallow Lands						
X_1	Slope						
X_2	Ruggedness Number						
X_3	Mean Annual Rainfall						
X_4	Drainage Density						
X_5	Quality of Water in *p*H						
X_6	Behaviour of Water Table	1.000					
X_7	Nitrogen of the Soil	—0.992***	1.000				
X_8	Phosphate of the Soil	—0.996***	0.990***	1.000			
X_9	Potash of the Soil	—0.986***	0.988***	0.987***	1.000		
X_{10}	*p*H Value of the Soil	0.975***	—0.973***	—0.981***	—0.992***	1.000	
X_{11}	Land Concentration	0.994***	—0.990***	—0.997***	—0.991***	0.987***	1.000

* Significant at 10% level of significance.
** Significant at 5% level of significance.
*** Significant at 1% level of significance.

correlation between the percentage of old fallow land and the quality of water in *p*H significant at 1% level of significance.

Most of the sample villages with very low (Table 3.10) water table have very high percentage of old fallow land because in the low water table areas the soil moisture is low and land is unfit for producing the crop because of which farmers leave such land uncultivated. Statistical correlation (Table 3.11) of these two variables provides enough clue that there is significant positive correlation significant at 1% level of significance.

3.5.4 Pedogenic (Nitrogen X_7, Phospate X_8, Potash X_9 and *p*H Value of the Soil X_{10})

It has been recorded from Table 3.10 that there is little variation among the sample villages in nitrogen, phosphate and potash of the soil and that is why these variables are not considered for further analysis.

A high pH value of the soil is harmful because it has more amount of salt which is a hindrance to crop production. It can be observed from Table 3.10 that the *p*H value of the soil is high in those villages where percentage of old fallow land is high, thereby indicating that there is a positive correlation between percentage of old fallow land and *p*H, value of the soil (Table 3.11) and the above relationship is significant at 1% level of significance.

3.5.5 Anthropogenic (Land Concentration/Gini's Coefficient)

Land concentration is recorded very high (0.62 to 0.67) in Raibigo, Bhojpur and Bhagupur (Table 3.10) villages having a high percentage of old fallow land and land concentration is recorded very low in those villages where the percentage of old fallow land is very low because the farmers who have large-sized land holdings are least bothered to improve such lands because they produce little and the farmers are not sure to get a good output. Thus, due to the little attention paid to these lands there is an increase in the areal extent of these lands in most of the villages. It is also proved statistically (Table 3.11) that there is significant positive correlation between the percentage of old

fallow land and land concentration, significant at 1% level of significance.

3.5.6 Stepwise Regression Analysis

In the correlation matrix various variables of morphometric, climatic, hydrologic, pedogenic and anthropogenic factors have been tested. It has been recorded from the correlation matrix that there is multi-collinearity among the independent variables and that is why in different steps, stepwise regression is attempted which is depicted in Table 3.12. Results of the stepwise regression and its analysis are as follows :

Variable X_6 (behaviour of water table), in the first step of stepwise regression analysis, in the first step enters and explains 99.2% variation of old fallow land. In the second step variable X_{10} (*p*H value of the soil) enters and jointly with variable X_6 explains 99.3% variation of old fallow land. In the third step X_3 (mean annual rainfall) and in the fourth step variable X_4 (drainage density) are incorporated but the explaining capacity of these two variables is knocked down because of the decrease in the amount of R^2. So the main variables in this set of stepwise regression are the behaviour of water table and *p*H value of the soil in explaining the variation of old fallow land.

In the second set of stepwise regression variable X_{11} (land concentration) enters in the first step and alone explains 98.7% variation of old fallow land, while in the second step variable X_{10} (pH value of the soil) jointly with variable X_{11} explains 98.9% variation of old fallow land. In the third step variable X_3 and in the fourth step variable X_5 are incorporated but the explaining power of these variables is knocked down because in these steps R^2 becomes constant. So the main variables, in this set of stepwise regression, are land concentration and pH value of the soil in explaining variation of the old fallow land.

Thus from the above discussion it is recorded that the most dominant variables in explaining the total variation of old fallow land are land concentration, behaviour of water table and *p*H value of soil which explain 99.3% total variation of old fallow land where F value is significant at 1% level of significance.

TABLE 3.12 : **Old Fallow Land : Results of the Stepwise Regression Analysis**

Variables	Intercept	Regression Coefficient	S.E.	T	R^2	Increase in R^2	$\bar{R}$	$\bar{R}^2$	F
1	2	3	4	5	6	7	8	9	10
Setp 1) (6	0.14929	0.006	0.000	28.181***	0.992	—	0.99191	0.98388	794.171***
Step 2) (6	—0.23938	0.005	0.001	5.320***	0.993	0.002	0.99195	0.98396	396.956***
) (10		0.054	0.054	0.990					
Step 3) (6	—0.23938	0.004	0.001	3.662***	0.993	0.000	0.99149	0.98305	248.604***
) (10		0.053	0.056	0.518					
) (4		0.003	0.005	0.518					
Step 4) (6	—0.23141	0.004	0.001	3.422***	0.993	0.000	0.99076	0.98161	170.471***
) (10		0.053	0.058	0.899					
) (4		0.003	0.006	0.510					
) (3		—0.000	0.000	—0.237					
Step 1) (11	1.38548	1.171	0.053	22.107***	0.987	—	0.98696	0.97409	488.721***
Step 2) (11	8.26455	1.679	0.316	5.312***	0.989	0.001	0.98850	0.97713	276.657***
) (10		—0.941	0.578	—1.627					

Step 3) (11	8.19652	1.655	0.335	4.939***	0.989	0.000	0.98768	0.97551	171.015***
) (10		—0.924	0.603	—1.533					
) (3		—0.000	0.001	—0.352					
Step 4) (11	8.48178	1.685	0.481	3.504***	0.989	0.000	0.98656	0.97330	116.699***
) (10		—0.927	0.633	—1.465					
) (3		—0.000	0.001	—0.345					
) (5		—0.035	0.390	—0.090					

* Significant at 10% level of significance.
** Significant at 5% level of significance.
*** Significant at 1% level of significance.

3.6 FACTORS IN FALLOW LAND

Fallow land includes those lands which are kept fallow for more than one year and not less than two years. The investigations from the revenue officials and villagers reveal that the problem of keeping such land fallow is mainly due to the deteriorating condition of the soil. Here, morphometric, climatic, hydrologic, pedogenic and anthropogenic factors are considered to explain the variations in fallow land.

The sample villages selected for analysing the responsible factors in fallow land on the basis of quartiling (Appendix 5) are : Very High percentage (32.88 to 34.00) villages are Bhadainya, Ahibaranpur and Rampur Jaria; High percentage (27.97 to 32.50) villages are Devaritaranpatti, Sewara and Lamkana Dubepur; Medium percentage (17.06 to 24.74) villages are Suraiya Puregaon, Amkol and Sakardepur; Low percentage (9.50 to 15.13) villages are Baithu, Kataka and Shehoria; and Very Low percentage (1.84 to 6.60) villages are Inayatpur, Karaute and Tirahi.

3.6.1 Morphometric (Slope X_1 and Ruggedness Number X_2)

Slope and ruggedness number under the morphometirc factor are recorded very high (Table 3.13) in those villages where the percentage of fallow land is very high, while very low percentage fallow land villages have very low slope and ruggedness number. It is recorded that the slope and ruggedness number are correlated while there is little variation in these two variables within the sample villages. The same reasoning as given in the previous analysis may account for this positive correlation due to which the moisture retaiining capacity of the soil decreases, which in turn is responsible for the deteriorating condition o1 the fertility of the soil. Due to the abovementioned problem farmers are bound to leave the land uncultivated. It is also proved from the correlation matrix (Table 3.14) that there is significant positive correlation, significant at 1% level of significance between the percentage of fallow land, slope and ruggedness number.

TABLE 3.13 : **Factors in Fallow Land, 1981-82**

Sample Villages	% of Fallow Land to Total area	Slope in degrees	Ruggedness number	Mean Annual Rainfall in mms.	Drainage Density	Quality of Water in *p*H	Behaviour of Water Table in feet	Soil				Land Concentration/ Gini's Coefficient
								Nitrogen	Phosphate	Potash	*p*H Value	
	(Y)	(X_1)	(X_2)	(X_3)	(X_4)	(X_5)	(X_6)	(X_7)	(X_8)	(X_9)	(X_{10})	(X_{11})
Bhadainya	34.00	1.68	0.19	213	6.32	7.59	14	1.34	1.82	2.67	7.86	0.72
Ahibaranpur	33.39	1.67	0.19	210	6.31	7.58	15	1.37	1.85	2.68	7.87	0.71
Rampur Jaria	32.88	1.65	0.19	215	6.30	7.57	19	1.40	1.87	2.79	7.85	0.70
Devaritaranpatti	32.50	1.61	0,18	220	6.25	7.56	18	1.45	1.90	2.85	7.84	0.68
Sewara	32.46	1.58	0.18	215	6.20	7.55	20	1.47	1.92	2.92	7.73	0.64
Lamkandubepur	27.97	1.52	0.17	225	6.18	7.54	22	1.51	1.95	3.10	7.72	0.61
Saraiyapuregaon	24.74	1.48	0.17	195	6.15	7.52	18	1.53	1.99	3.18	7.71	0.58
Amkol	20.83	1.39	0.16	190	6.10	7.51	15	1.57	2.07	3.27	7.69	0.52
Sakardepur	17.06	1.32	0.15	195	6.05	7.46	18	1.60	2.09	3.35	7.68	0.48
Baithu	15.13	1.26	0.14	190	6.00	7.45	20	1.67	2.13	3.41	7.66	0.45
Kataka	12.65	1.20	0.13	187	5.85	7.47	15	1.69	2.17	3.62	7.58	0.38
Shehoria	9.50	1.15	0.12	182	5.70	7.44	17	1.75	2.21	3.75	7.55	0.32
Inayatpur	6.60	1.07	0.10	180	5.65	7.38	13	1.78	2.25	3.91	7.45	0.27
Karaute	3.75	1.00	0.09	175	5.58	7.33	12	1.85	2.28	4.76	7.46	0.18
Tirahi	1.84	0.92	0.07	195	5.45	7.31	10	1.92	2.33	4.88	7.41	0.11

TABLE 3.14 : **Correlation Matrix**

Sl. No.	Variables	Y	X_1	X_2	X_3	X_4	X_5
1	2	3	4	5	6	7	8
Y	% of the Fallow Lands	1.000					
X_1	Slope	0.996***	1.000				
X_2	Ruggedness Number	0.977***	0.988***	1.000			
X_3	Mean Annual Rainfall	0.933***	0.927***	9.22***	1.000		
X_4	Drainage Density	0.971***	0.982***	0.993***	0.923***	1.000	
X_5	Quality of Water in *p*H	9.969***	0.979***	0.986***	0.919***	0.969***	1.000
X_6	Behaviour of Water Table	0.572**	0.572**	0.635**	0.658**	0.628**	0.601**
X_7	Nitrogen of the Soil	—0.984***	—0.994***	—0.986***	—0.909***	—0.982***	—0.778***
X_8	Phosphate of the Soil	—0.994***	—0.996***	—0.975***	—0.914***	—0.972***	—0.966***
X_9	Potash of the Soil	—0.938***	—0.955***	—0.974***	—0.890***	—0.964***	—0.977***
X_{10}	*p*H Value of the Soil	0.982***	0.985***	0.981***	0.901***	0.978***	0.962***
X_{11}	Land Concentration	0.987***	0.995***	0.995***	0.929***	0.993***	0.984***

Sl. No.	Variables	X_6	X_7	X_8	X_9	X_{10}	X_{11}
		9	10	11	12	13	14
Y	% of the Fallow Lands						
X^1	Slope						
X_2	Ruggedness Number						
X_3	Mean Annual Rainfall						
X_4	Drainage Density						
X_5	Quality f Water in pH						
X_6	Behaviour of Water Table	1.000					
X_7	Nitrogen of the Soil	−0.528**	1.000				
X_8	Phosphate of the Soil	−0.538**	0.992***	1.000			
X_9	Potash of the Soil	−0.644**	0.962***	0.938***	1.000		
X_{10}	pH Value of the Soil	0.578**	−0.980***	−0.980***	−0.946***	1.000	
X_{11}	Land Concentration	0.615**	−0.992***	−0.987***	−0.975***	0.984***	1.000

* Significant at 10% level of significance.
** Significant at 5% level of significance.
*** Significant at 1% level of significance.

3.6.2 Climatic (Mean Annual Rainfall X_3)

Most of the sample villages suffer much from the high amount of rainfall (Table 3.13) and these are the villages which have a high percentage of fallow land, and low mean annual rainfall is found in those sample villages which have a low percentage of fallow and. In most of these villages it is observed that in the rainy season due to high rainfall, kharif cropping is not possible because of the flooding as the affected areas are mainly surrounded by rail, road, in a few cases by canal which obstruct the outflow of the water. After the rainy season when water dries up, the same land gives rise to weed infestation which is another hindrance to crop production. Due to the reasons cited above land is left uncultivated, which is proved from the correlation matrix (Table 3.14), that there is a significant positive correlation between these two variables significant at 1% level of significance.

3.6.3 Hydrologic (Drainage Density X_4, Quality of Water in *p*H X_5 and Behaviour of Water Table X_6)

The drainage lines distribute themselves in varied fashions following the structural alignments. In this region frequency of occurrence and density of drainage lines ranges from a meagre value to higher value over the rugged areas. The drainage lines which have been taken into account have a consideration of both the black and blue lines. The high drainage density gives rise to higher dissection of the land whereby the top soils of the strata do not remain in equilibrium which makes the land uncultivable. It can also be seen from Table 3.13, that the high drainage density is found in those villages where the percentage of fallow land is high and it is low in those villages where the percentage of fallow land is low. It is proved from Table 3.14 that there is significant positive correlation between fallow land and drainage density, significant at 1% level of singnificance.

Low quality of water in pH provides more scope for the formation of fallow land because when this type of brackish water is used for irrigational purposes it spreads salts in the field. According to the field observation in most of the areas salt has disturbed the fertility status of the soil because of which most of these lands are left fallow. In the sample villages it is observed that the villages having high quality water in pH (Table 3.13) also have high percentage of fallow land, and the quality of water in *p*H is low in those villages where the percentage of fallow land is also low. There is a significant positive correlation (Table 3.14) significant at 1% level of significance between fallow land and quality of water in *p*H.

It it recorded from Table 3.13 that Inayatpur, Karaute and Tirahi (very low percentage fallow land villages) have a very high water table and the water table is found low in Sewara, Baithur and Lamkana, Dubepur villages. Villages with a high water table have already enough moisture in the soils, and in the rainy season there is added accumulated water and this results in logging conditions because of which the farmers leave the land uncultivated. In the summer season the area is affected by weed infestation because of the prevailing moisture. Villages with a low water table are characterised by less moisture and by this fertility status of the soil gets affected and the farmers leave the land fallow. Thus there are both these types of cases prevalent in Sultanpur District. It is observed from Table 3.14 that there is significant positive correlation between the behaviour of water table and percentage of fallow land, significant at 5% level of significance. Thus the assumed hypothesis is not strongly supported statistically while the reasons for this relationship are discussed above.

3.6.4 Pedogenic (Nitrogen X_7, Phosphate X_8, Potash X_9 and *p*H Value of the Soil X_{10})

Nitrogen, phosphate and potash of the soil have not been considered for further analysis because there is little variation among the sample villages and these variables are taken only to know the fertilizer requirement of the soil to produce the crop.

It is recorded from Table 3.5 that the pH value of the soil is very high in those villages where the percentage of fallow land is high and it is low in low percentage fallow land sample villages because the soil with a high pH value has excessive amount of salt which is a hindrance to crop production. It is proved from the correlation matrix (Table 3.14) that there is significant positive correlation between pH value of the soil and percentage of fallow land significant at 1% level of significance.

3.6.5 Anthropogenic (Land Concentration/Gini's Co-efficient X_{11})

Land concentration is recorded very high (0.70 to 0.72) in Bhadainya, Ahibaranpur and Rampur Jaria, as given in Table 3.13 and it is very low (0.11 to 0.27) in Inayatpur, Karaute and Tirahi villages, where the percentage of fallow land is also low. There is positive significant correlation at 1% level of significance between percentage of fallow land and land concentration because the owners of large size land holdings pay less attention to these lands where fertility of the soil is less. Instead of improving these lands, they leave them fallow because they require more input, which the farmers are not ready to pay for. On the other hand, they divert their resources to the good land which requires less attention compared to the former.

3.6.6 Stepwise Regression Analysis

The system of explanation by X independent variable discussed above shows a high degree of multi-collinearity, among themselves. To remove inconsistencies arising out of this problem of multi-collinearity on statistical exercise is attempted which is depicted in Table 3 15.

Stepwise regression analysis is attempted taking fallow land as dependent variable (Y) and other environmental and anthropogenic factors as independent variables (X). The stepwise regression analysis attempted in two sets of stepwise regression gives an intercorrelation matrix which is given in Table 3 15.

The first set of stepwise regression shows that in the first step variable X_4 (drainage density) enters where R^2 is found to be 0.993. Thus the drainage density explains 99.3% variation of fallow land. In the second step variable X_1O (pH value of the soil) jointly with variable X_4 explains 99.4% total variation of fallow land. In the third step variable X_6 (behaviour of water table) and in the fourth step X_3 (mean annual rainfall) are incorporated but in these steps R^2 starts declining and therefore

TABLE 3.15 : **Fallow Land : Results of the Stepwise Regression Analysis**

Variables	Intercept	Regression coefficient	S.E.	T	R^2	Increase in R^2	$\bar{R}$	$\bar{R}^2$	F
1	2	3	4	5	6	7	8	9	10
Step 1) (4	—0.65667	0.134	0.004	29.952***	0.993	—	0.99283	0.98571	987.139***
Step 2) (4	—0.89979	0.105	0.021	5.097***	0.994	0.001	0.99342	0.98688	486.811***
) (10		0.054	0.038	1.442					
Step 3) (4	—0.91945	0.099	0.022	4.387***	0.994	0.000	0.99327	0.98658	315.031***
) (10		0.061	0.039	1.560					
) (6		0.000	0.000	0.800					
Step 4) (4	—0.91120	0.096	0.025	3.799**=	0.994	0.000	0.99270	0.98545	216.155***
) (10		0.061	0.041	1.481					
) 6		0.000	0.001	0.671					
) (3		0.000	0.000	0.250					
Step 1) (11	0.74045	1.278	0.036	35.662***	0.995	—	0.99493	0.98988	1271.767***
Step 2) (11	–1.59455	1.029	0.196	5.258***	0.996	0.001	0.99621	0.99044	669.583***
) (10		0.320	0.248	1.293					

TABLE 3.15 (*Contd.*)

1	2	3	4	5	6	7	8	9	10
Step 3) (11	—1.86165	0.965	0.242	3.981***	0.996	0.000	0.99491	0.98984	417.957***
) (10		0.345	0.261	1.323					
) (3		0.001	0.001	0.483					
Step 4) (11	—2.28703	0.937	0.357	2.622***	0.996	0.000	0.99446	0.98895	295.310***
) (10		0.351	0.279	1.260					
) (3		0.001	0.001	0.457					
) (5		0.053	0.488	0.109					

* Significant at 10% level of significance.
** Significant at 5% level of significance.
*** Significant at 1% level of significance.

the analysis at these steps is truncated and these variables have no explaining power of the dependent variable.

The most dominant variable in explaining variability of fallow land, in the second set of stepwise regression, is variable X_{11} (land concentration) which alone explains 99.5% total variation of fallow land and its degree of freedom is significant at 1% level of significance. In the second step variable X_{10} (pH value of the soil) jointly with variable X_{11} explains 99.6% variation of fallow land. But in the third step variable X_3 (mean annual rainfall) and in the fourth step variable X_5 (quality of water in pH) could not come out with a strong correlation.

Thus the hypothesis related to this empirical exercise is supported by the above mentioned variables. The most important variables in explaining the variation of fallow land in this case study are land concentration, drainage density and pH value of the soil, while the other variables are not found on the optimal regression line.

3.7 FACTORS IN KANKRILI LAND

The kankrili land of the area has calcium carbonate nodules called kankar, often occurring at about a depth of one meter in which excessive amount of soluble sodium is recorded. This inturn determines high pH value of the soil. After testing of the soil, it is found that there is deficiency of available calcium, nitrogen and micro nutrients like zinc and of nutritional disorder. There is an extremely low water intake rate and hydraulic conductivity resulting in the present analysis morphometric, climatic, hydrologic and pedogenic factors are taken to analyse the variation of kankrili land.

The sample villages for analysing the responsible factors in kankrili land on the basis of stratified sampling are: Very High percentage (12.98 to 19.50) villages are Seurchamurkha, Karamchandrapur and Babarpur; High percentage (8.22 to 11.69) villages are Ardih, Guregaon and Samsapur; Medium percentage (4.20 to 7.47) villages are Sathari, Nagaipur and Payagpatti; Low percentage (2.24 to 3.83) villages are Ajhui, Karaundipur and Saifullaganj; and Very Low percentage (0.53 to 1.22) villages are Shehoria, Basauthi and Sarwan.

3.7.1 Morphometric (Slope X_1 and Ruggedness Number X_2)

Slope and ruggedness number of the morphometric factors are recorded very high (Table 3.16) in very high percentage kankrili land sample villages and it is very low in very low percentage of kankrili land sample villages. In the sample villages, it is observed that the general topography itself is undulating on which the kankar formation has taken place. Table 3.17 provides enough clue that there is significant positive correlation between percentage of kankrili land and slope and ruggedness number significant at 1% level of significance.

3.7.2 Climatic (Mean Annual Rainfall X_3)

The independent variable mean annual rainfall is not considered for further analysis because there is low variation in mean annual rainfall at village level (Table 3.16) and infromation for the above variable is generalised from the tahsil and block level information. It is also clear from the correlation matrix (Table 3.17) [that there is insignificant correlation between percentage of kankrili land and mean annual rainfall.

3.7.3 Hydrologic Drainage (Density X_4, Quality of Water in *p*H X_5 and Behaviour of Water Table X_6)

Drainage density is recorded very high in very high kankrili land sample villages and very low in low percentage of kankrili land villages (Table 3.16) because a high degree of artificial and natural overland flow leads to more erosion due to which little abraded kankar calcium nodules are deposited down from higher to lower areas and due to this kankrili land in many villages is recorded to be increasing. It is statistically proved from Table 3.17 that there is significant positive correlation between percentage of kankrili land and drainage density, significant at 1% level of a significance.

It is recorded from Table 3.16 that the quality of water in pH has the same distribution pattern as it is in the case of kankrili land among the sample villages. Villages having a high percentage of kankrili land also have a high quality of water in pH because in the kankrili land there is already an excessive

TABLE 3.16 : Factors in Kankrili Land, 1981-82

Sample Villages	% of Kankrili land to total area	Slope in degrees	Ruggedness Number	Mean Annual Rainfall in mms.	Drainage Density	Quality of water in pH	Behaviour of water table in feet	Soil: Nitrogen	Soil: Phosphate	Soil: Potash	Soil: pH Value	Depth of the base rock in feet
	Y	X_1	X_2	X_3	X_4	X_5	X_6	X_7	X_8	X_9	X_{10}	X_{11}
Seurchamarkha	19.50	7.13	0.74	169	5.85	7.87	17	2.06	1.67	3.11	8.08	5.8
Karamchandrapur	13.40	7.06	0.71	158	5.82	7.83	15	2.09	1.68	3.08	8.05	5.5
Babarpur	12.98	6.78	0.67	215	5.73	7.82	14	2.12	1.75	3.02	7.96	5.1
Ardih	11.69	6.53	0.66	195	5.61	7.78	15	2.18	1.85	2.95	7.85	4.8
Guregaon	11.12	6.37	0.64	210	5.57	7.86	14	2.27	1.92	2.87	7.84	4.5
Samsapur	8.22	6.22	0.62	205	5.42	7.75	13	2.35	1.98	2.75	7.76	4.0
Sathari	7.47	5.93	0.60	190	5.32	7.71	12	2.42	2.07	2.68	7.73	3.8
Nagaipur	5.77	5.84	0.59	185	5.19	7.67	11	2.51	2.16	2.62	7.72	3.5
Payagpatti	4.20	5.45	0.55	178	5.06	7.65	10	2.58	2.25	2.58	7.68	3.2
Ajhui	3.83	5.17	0.52	165	4.84	7.63	11	2.67	2.38	2.55	7.65	3.0
Karaundipur	3.77	4.76	0.48	185	4.67	7.54	10	2.79	2.49	2.50	7.64	2.8
Saifullaganj	2.24	3.47	0.35	175	4.78	7.52	12	2.85	2.58	2.48	7.55	2.4
Shehoria	1.22	3.12	0.31	182	4.52	7.50	10	2.98	2.65	2.41	7.51	2.0
Basauthi	0.86	2.78	0.28	195	4.38	7.48	8	3.06	2.78	2.38	7.46	1.8
Sarwan	0.53	1.42	0.15	190	3.86	7.43	9	3.12	2.92	2.27	7.42	1.5

TABLE 3.17 : **Correlation Matrix**

Sl. No.	Variables	Y	X1	X_2	X_3	X_4	X_5
Y	% of Kankrili Lands	1.000					
X_1	Slope	0.862***	1.000				
X_2	Ruggedness Number	0.872***	0.999***	1.000			
X_3	Mean Annual Rainfall	0.124	0.133	0.111	1.000		
X_4	Drainage Density	0.911***	0.960***	0.969***	0.192	1.000	
X_5	Quality of Water in pH	0.933***	0.948***	0.949***	0.259	0.970***	1.000
X_6	Behaviour of Water Table	0.946***	0.815***	0.827***	0.090	0.895***	0.893***
X_7	Nitrogen of the Soil	—0.941***	—0.964***	—0.966***	—0.200	—0.982***	—0.982***
X_8	Phosphate of the Soil	–0.938***	—0.970***	—0.970***	—0.197	—0.989***	—0.982***
X_9	Potash of the Soil	0.972***	0.911***	0.915***	0.148	0.961***	0.958***
X_{10}	pH Value of the Soil	0.969***	0.927***	0.932***	0.044	0.950***	0.948***
X_{11}	Depth of Base Rocks	0.973***	0.941***	0.945***	0.125	0.969***	0.971***

Sl. No.	Variables	X_6	X_7	X_8	X_9	$X1_0$	X_{11}
Y	% of Kankrili Lands						
X1	Slope						
X_2	Ruggedness Number						
X_3	Mean Annual Rainfall						
X_4	Drainage Density						
X_5	Quality of Water in pH						
X_6	Behaviour of the Water Table	1.000					
X_7	Nitrogen of the Soil	—0.910***	1.000				
X_8	Phosphate of the Soil	—0.898***	0.997***	1.000			
X_9	Potash of the Soil	0.941***	—0.976***	—0.976***	1.000		
XI_0	pH Value of the Soil	0.916***	—0.970***	—0.977***	0.985***	1.000	
X_{11}	Depth of Base Rocks	0.933***	—0.989***	—0.988***	0.993***	0.992***	1.000

* Significant at 10% level of significance.
** Significant at 5% level of significance.
*** Significant at 1% level of significance.

amount of exchangeable soluble sodium salt and if high quality of underground water in pH is used for irrigation in the adjoining areas, it will also lead to the formation of kankar nodules. It can be seen from Table 3.17 that there is significant positive correlation between percentage of kankrili land and quality of water in pH significant at 1% level of significance.

The water table is recorded low in very high percentage of kankrili land sample villages and vice versa. Because in these areas the nature of the soil is hard and there is less possibility of leaching/percolation of water, underground water is found below the hard-pan strata of the soil. It is proved from Table 3.17 that there is positive significant correlation between percentage of kankrili land and behaviour of water table, significant at 1% level of significance.

3.7.4 Pedogenic (Nitrogen X_7, Phosphate X_8, Potash X_9, pH Value of the Soil X_{10} and Depth of Base Rock X_{11})

Nitrogen, phosphate and potash of the soil are not considered for further analysis because there is little variation in these variables (Table 3.16) and these were taken mainly to know the fertility requirement of the soil to produce certain crops.

The pH value of the soil is recorded high in high percentage of kankrili land sample villages and the pH value of the soil is low (Table 3.16) in low percentage of kankrili land sample villages because of the existing excessive amount of exchangeable soluble sodium salt which is a hindrance to crop production. It is proved statistically (Table 3.17) that there is significant positive correlation between percentage of kankrili land and pH value of the soil, significant at 1% level of significance.

Depth of the base rock is found very high (Table 3.16) in those villages where the percentage of kankrili land is high and it is low in low percentage kankrili land sample villages because in such type of land there is more presence of calcium carbonate nodules and there is hard kankar pan strata at a lower depth. It is recorded low in lower areas where formation of kankar has just started, i.e. calcium carbonate nodules due to high soluble sodium salt. It has been proved from the correlation matrix that there is significant positive correlation between percentage of kankrili land and depth of base rock, significant at 1% level of significance.

3.7.5 Stepwise Regression Analysis

A stepwise regression analysis is attempted taking the percentage of kankrili land as dependent variable (Y) and morphometric, climatic, hydrologic and pedogenic factors as independent variables (X). The stepwise regression analysis provides an inter-correlation matrix which is given in Table 3.18 in three sets of stepwise regression. In all sets of stepwise regression analysis at certain steps R^2 starts declining and analysis gets truncated at such steps.

In the first set of stepwise regression, dominant variable is X_{10} (pH value of the soil) which alone explains 92.7% variation of kankrili land. In the second step variable X_4 (drainage density) enters and jointly with variable X_{10} explains 93.1% variation of kankrili land, while in the third step variable X_6 (behaviour of water table) enters and jointly with variable X_{10} and X_4 explains 93.6% total variation in kankrili land.

In the second set of stepwise regression analysis, in the first step variable X_5 (quality of water in pH) enters and explains 94.9% variation of kankrili land while in the second step variable X_4 (drainage density) with variable X_5 (quality of water in pH) explains 95.9% variation of kankrili land. Lastly in the third step variable X_{11} (depth of base rock) is incorporated which jointly with variables X_4 and X_5 explains 96.0% of total variation in kankrili land.

In the third set of stepwise regression analysis, the dominant variable is X_4 (drainage density) which explains 19.2% total variation of kankrili land and in the second step variable X_6 (behaviour of water table) is incorporated which jointly with variable X_4 explains 26.5% total variation of the kankrili land.

Thus the high degree of multi-collinearity among explanatory variable suggest that the effect of all variables in all the sets of stepwise regression is subsumed. Thus the main important variables in explaining total variation of kankrili land are pH value of the soil, quality of water in pH, behaviour of water table, depth of base rock and drainage density.

TABLE 3.18 : **Kankrili Land :**
Results of the Stepwise Regression Analysis

Variables	Intercept	Regression coefficient	S.E.	T	R^2	Increase in R^2	$\bar{R}$	$\bar{R}^2$	F
Step 1) (10	—56.54288	7.991	0.898	8.896***	0.927	—	0.92677	0.85890	79.133***
Step 2) (10	—58.07806	7.956	0.907	8.755***	0.931	0.004	0.92554	0.85614	39.300***
) (4		0.010	0.011	0.885					
Step 3) (10	—70 92189	9.844	2.282	4.314***	0.936	0.005	0.92528	0.85614	26.069***
) (4		0 011	0.011	0.984					
) (6		—0.164	0.182	—0.903					
Step 1) (5	— 8.05847	1.119	0.103	10.885***	0.949	—	0.94927	0.90111	118.474***
Step 2) (5	— 8.11486	1.163	0.009	11.706***	0.959	0.010	0.95622	0.91435	69.442***
) (4		—0.002	0.001	—1.709*					
Step 3) (5	— 6.89065	0.990	0.504	1.963*	0.960	0.001	0.95302	0.90824	42.954***
) (4		—0.001	0.001	—1.196					
) (11		0.019	0.054	0.351					
Step 1) (4	157.07765	5.401	7.664	0.705	0.192	—	0.19183	0.03679	19.906***
Step 2) (4	136.56407	15.716	17.556	0.895	0.265	0.073	0.036280	0.00131	17.491***
) (6		—2.667	4.061	—0.659					

* Significant at 10% level of significance
** Significant at 5% level of significance
*** Significant at 1% level of significance

3.8 FACTORS IN RAVINE LAND

The broad literature survey which is discussed in the first chapter provides that there is much work done by various academic bodies and government institutions on distribution of ravine lands in India, where they have discussed various methods and economics of its reclamation but less attention has been paid to an analysis of the factors influencing ravine land. From the observations of the field, it is clear that the environmental factors are the main responsible factors influencing ravine land in which for the present analysis mainly various varibales of morphometric, climatic, hydrologic and pedogenic are taken into consideration.

The 15 sample villages selected for analysing the factors influencing ravine land on the basis of quartiling as briefed in the first chapter (Appendix 7) are: Very High percentage (20.37 to 35.31) villages are Ratanpur, Khajapur and Upadhyapur; High percentage (16.01 to 17.30) villages are Ardih Kailashpur and Payagapatti; Medium percentage (10.90 to 15.48) villages are Kumhai, Semari Dargar and Chiranedih; Low percentage (5.56 to 9) villages are Grantkurwar, Karaundipur and Sherpur; and Very Low percentage (1.18 to 3.21) villages are Saidpur, Sarwan and Chandpur Sadar.

3.8.1 Morphometric (Slope X_1 and Ruggedness Number X_2)

The variables slope and ruggedness number are found to be very high in very high percentage of ravine land sample villages and these variables are very low in very low percentage of ravine land villages because in such ravine areas, greater relative heights give rise to high slope and ruggedness number. Where the formation of a ravine has just started, the same values are recorded low. It is proved from the correlation matrix (Table 3 20) that there is positive correlation between slope and ruggedness number and percentage of ravine land which are significant at 1% level of significance.

3.8.2 Climatic (Mean Annual Rainfall X_3)

Mean annual rainfall has little variation among the sample

villages, but rainfall in the area does affect the ravine through erosion. It is recorded in the rainy season that most of the ravinous areas of the villages are eroded through rainwater where deepening and widening of the ravines increases in each successive rainy season. The data for the mean annual rainfall of the sample villages are generalied from tahsil and block level information as the village level data was not available. It can be seen from Table 3.20 that there is negative correlation between mean annual rainfall and percentage of ravine land which is significant at 10% level of significance.

3.8.3 Hydrologic (Drainage Density X_4, Quality of Water in pH X_5 and Behaviour of Water Table X_6)

Drainage density among the sample villages is found very high (Table 3.19) in very high percentage of ravine land sample villages and it is low in low percentage of ravine land sample villages. This is due to more dissection in the form of higher drainage density which ultimately leads to more ruggedness. The correlation matrix (Table 3.20) provides enough clue that there is significant positive correlation between percentage of ravine land and drainage density significant at 1% level of significance.

High quality of water in pH is observed in those villages where the percentage of ravine land is high and it is low in low percentage of ravine land sample villages (Table 3.19). Though this factor is not of much significance as far as ravine formation is concerned but, later on, the utilisation of the same may well affect the area. The high pH value of the water is attributed to transportation of the salts in the deeps of ravines by the surface wash. From the table it is clear that though there is decrease in the value of quality of water in pH the decrease is nominal and also ravine statistics show a marked variation within the sample villages. This may be the reason for the significant positive correlation within these variables. It is also proved statistically (Table 3.20) that there is positive significant correlation at 1% level of significance between percentage of ravine land and quality of water in pH.

Low water table in ravine land is recorded where percentage of ravine land is high due to higher ruggedness and water table

TABLE 3.19 : Factors in Ravine Land, 1981-82

	% of Ravine land to total area	Slope in degrees	Ruggedness Number	Mean Annual Rainfall in mms.	Drainage Density	Quality of water in pH	Behaviour of water table in feet	Soil			
								Nitrogen	Phosphate	Potash	pH Value
	Y	X_1	X_2	X_3	X_4	X_5	X_6	X_7	X_8	X_9	X_{10}
Ratanpur	35.31	9.39	0.99	185	5.20	7.67	29	1.99	1.70	2.13	7.67
Khajapur	24.35	9.35	0.93	170	5.15	7.65	27	2.01	1.72	2.15	7.65
Upadhyapur	20.37	9.23	0.92	175	5.02	7.63	26	2.05	1.75	2.18	7.62
Ardih	19.28	9.15	0.91	195	4.91	7.62	25	2.08	1.78	2.23	7.59
Kailashpur	17.30	9.04	0.90	165	4.82	7.61	24	2.12	1.82	2.29	7.54
Payagpatti	16.01	8.92	0.89	178	4.70	7.60	22	2.18	1.87	2.36	7.52
Kumhai	15.48	8.81	0.88	158	4.56	7.65	21	2.27	1.90	2.46	7.50
Semari Qargar	14.90	8.61	0.86	178	4.38	7.66	20	2.39	1.92	2.47	7.46
Chiranedih	10.90	8.48	0.84	185	4.17	7.57	19	2.47	1.99	2.68	7.43
Grantkurwar	8.00	8.27	0.82	168	3.92	7.54	18	2.52	2.05	2.75	7.42
Karondipur	7.17	8.12	0.81	185	3.84	7.56	17	2.68	2.09	2.88	7.40
Sherpur	5.56	7.85	0.78	182	3.71	7.53	16	2.78	2.15	2.97	7.38
Saidpur	3.21	7.64	0.76	186	3.54	7.52	15	2.81	2.27	3.12	7.33
Sarwan	2.32	7.41	0.74	190	2.78	7.46	13	3.15	2.35	3.21	7.30
Chandpur Sadar	1.18	7.12	0.71	195	2.41	7.41	10	3.32	2.47	3.35	7.23

TABLE 3.20 : **Correlation Matrix**

Sl. No.	Variables	Y	X_1	X_2	X_3	X_4	X_5
Y	% of the Ravine Lands	1.000					
X_1	Slope	0.921***	1.000				
X_2	Ruggedness Number	0.969***	0.985***	1.000			
X_3	Mean Annual Rainfall	—0.366*	—0.627**	—0.543**	1.000		
X_4	Drainage Density	0.896***	0.987***	0.970***	—0.640**	1.000	
X_5	Quality of Water in pH	0.856***	0.927***	0.920***	—0.617**	0.946***	1.000
X_6	Behaviour of Water Table	0.945***	0.987***	0.985***	—0.543**	0.978***	0.962***
X_7	Nitrogen of the Soil	—0.890***	—0.992***	—0.970***	0 667**	—0.996***	—0.933***
X_8	Phosphate of the Soil	—0.912***	—0.966***	—0.981***	0.624**	—0.992***	—0.942***
X_9	Potash of the Soil	—0.919***	—0.996***	—0.980***	0.631**	—0.980***	—0.930***
X_{10}	pH Value of the Soil	0.951***	0.982***	0.984***	—0.539**	0.971***	0.987***

Sl. No.	Variables	X_6	X_7	X_8	X_9	X_{10}
Y	% of the Ravine Lands					
X_1	Slope					
X_2	Ruggedness Number					
X_3	Mean Annual Rainfall					
X_4	Drainage Density					
X_5	Quality of Water in pH					
X_6	Behaviour of Water Table	1.000				
X_7	Nitrogen of the Soil	—0.977***	1.000			
X_8	Phosphate of the Soil	—0.984***	0.991	1.000		
X_9	Potash of the Soil	—0.981***	—0.986***	0.992***	1.000	
X_{10}	pH Value of the Soil	0.996***	—0.967***	—0.976***	—0.976***	1.000

* Significant at 10% level of significance.
** Significant at 5% level of significance.
*** Significant at 1% level of significance.

is high (Table 3.19) in low percentage of ravine land sample villages. It is also proved from the correlation matrix (Table 3.20) that there is significant positive correlation between percentage of ravine land and behaviour of water table significant at 1% level of significance.

3.8.4 Pedogenic (Nitrogen X_7, Phosphate X_8, Potash X_9 and pH Value of the Soil X_{10})

Nitrogen, phosphate, potash and pH value of the soil are not considered for further analysis because there is little variation of these variables within the sample villages.

High pH value of the soil (Table 3.19) is found in high percentage of ravine land sample villages and pH value of the soil is observed low in low percentage of ravine land sample villages. This may be attributed to the same reason as cited under the quality of water and its relation with ravines. It is proved statistically that there is significant positive correlation (Table 3.20) between percentage of ravine land and pH value of the soil, significant at 1% level of significance.

3.8.5 Stepwise Regression Analysis

The stepwise regression analysis is attempted taking percentage of ravine land as dependent variable (Y) and environmental factors as independent variable (X). Two sets of stepwise regression analysis are attempted to remove inconsistencies of multi-collinearity. In these sets of stepwise regression at certain steps R^2 starts declining due to the multi-collinearity among the exogenous variables as given in Table 3.21. Therefore the analysis is truncated at these steps. The variables recorded on the optimal regression line in different sets of stepwise regression analysis are :

In the first set of stepwise regression in the first step variable X_4 (drainage density) enters and explains 98.7% variation of ravine land. While in the second step variable X_6 (behaviour of water table) jointly with variable X_4 explains 99.2% variation of ravine land and in the third step X_1

TABLE 3.21 : **Ravine Lands : Results of the Stepwise Regression Analysis.**

Variables	Intercept	Regression coefficient	S.E.	T	R^2	Increase in R^2	$\overline{R}$	$\overline{R^2}$	F
Step 1) (4	4.89208	0.856	0.038	22.432***	0.987	—	0.98733	0.97482	503.189***
Step 2) (4	5.28902	0.444	0.148	3.004**	0.992	0.005	0.99189	0.98384	394.048***
) (6		0.067	0.023	2.855**					
Step 3) (4	6.62233	0.261	0.176	1.479	0.994	0.002	0.99297	0.98598	301.848***
) (6		0.089	0.026	3.472***					
) (3		—0.006	0.003	—1.657					
Step 1) (10	—3.58624	0.594	0.030	19.956***	0.984	—	0.98407	0.96839	398.242***
Step 2) (10	—4.32109	0.489	0.061	7.981***	0.988	0.004	0.98684	0.97385	241.146***
) (5		0.200	0 105	1.904*					
Step 3) (10	—4.45727	0.488	0.064	7.666***	0.988	0.0C0	0.98591	0.97202	149.173***
) (5		0.215	0.017	1.842*					
) (3		0.000	0.000	0.363					

* Significant at 10% level of significance

** Significant at 5% level of significance

*** Significant at 1% level of significance

(slope) is incorporated which explains, jointly with variable X_4 and X_6, 99.4% variation of ravine land.

The second set of stepwise regression analysis shows that in the first step variable X_{10} (pH value of the soil) enters which explains 98.4% variation of ravine land and in the second step variable X_5 (quality of water in pH) jointly with variable X_{10} explains 98.8% total variation of ravine land. In the third step variable X_3 (mean annual rainfall) enters and at this step R^2 declines because of which this variable has no explaining power in explaining the variation of ravine land.

Therefore, it can be concluded that the hypothesis related to abovementioned environmental factors with ravine land is supported by this empirical exercise. The high degree of multi-collinearity among the environmental factors suggests that the effect of various variables is subsumed. The dominant variables in explaining the variation of ravine land are drainage density, pH value of the soil, quality of water in pH, behaviour of water table and slope followed by mean annual rainfall.

3.9 FACTORS IN OTHER TYPES OF WASTE LAND

The other types of waste lands are those types of land which are not included in any of the abovementioned categories but which are temporarily left out of cultivation. To analyse the factors responsible for the formation of other types of waste land environmental (morphometric, climatic, hydrologic and pedogenic) and anthropogenic (land concentration) factors are taken into consideration.

The selected sample villages for analysing the influencing factors of other types of waste land, which are given in Appendix 8, are: Very High percentage (29.40 to 32.08) villages are Dadara, Kundabhairopur and Madanpur Devrar; High percentage (24.46 to 28) villages are Ranjitpur, Katgharakhurd and Rampurustada; Medium percentage (17.70 to 21.60) villages are Ramnagar, Chaubepur and Amkol; Low percentage (8.33 to 15.83) villages are Kunwarpur-I. Ballipur and Semari Dargar; and Very Low percentage (1.08 to 5.76) villages are Saifullaganj, Karondipur and Basauthi.

3.9.1 Morphometric (Slope X_1 and Ruggedness Number X_2)

The slope and ruggedness number of the morphometric factors are recorded very high (Table 3.22) in those sample villages where the percentage of other types of waste land is also high and these variables are low in low percentage of other types of waste land sample villages because there is more occurrence of soil erosion, which leads to loss of fertility of soil due to which the cultivation is not possible on these lands. It is statistically proved (Table 3.23) that there is significant positive correlation between percentage of other types of waste land and slope and ruggedness number, significant at 1% level of significance.

3.9.2 Climatic (Mean Annual Rainfall X_3)

Since the data for mean annual rainfall for the sample villages is not collected it is generalised from the tahsil and block level information. There is little variation of mean annual rainfall (Table 3.22) among sample villages and that is why this variable has been given less importance but the rainfall does affect the area through erosion, which in turn deteriorates the fertility status of the soil. It can be observed from Table 3.23 that there is insignificant positive correlation between mean annual rainfall and percentage of other types of waste land.

3.9.3 Hydrologic (Drainage Density X_4, Quality of Water in pH X_5 and Behaviour of Water Table X_6)

Drainage density is recorded very high (Table 3.22) in those villages where percentage of other types of waste land is high and drainage density is low in low percentage of other types of waste land sample villages because high drainage density (both natural and artificial) causes more erosion through which fertility of the soil lessens and farmers are bound to leave the land uncultivated. The above assumption is statistically (Table 3.23) supported, that there is significant positive correlation between percentage of other types of waste land and drainage density significant at 1% level of significance.

Quality of water in pH is observed very high (7.76% to 7.83%) in Dadara, Kundabhairopur and Madanpur Devrar

TABLE 3.22 : **Factors in other Types of Waste Land, 1981-82**

Sample Villages	% of other types of Waste Land to Total Area	Slope in Degrees	Ruggedness number	Mean Annual Rain-fall in mm.	Drainage Density	Quality of Water in pH	Behaviour of Water Table in feet	Soil: Nitrogen	Soil: Phosphate	Soil: Potash	Soil: pH Value	Land Concentration/ Gini's coefficient
	Y	X_1	X_2	X_3	X_4	X_5	X_6	X_7	X_8	X_9	$X_{,0}$	X_{11}
1	2	3	4	5	6	7	8	9	10	11	12	13
Dadara	32.08	3.12	0.32	158	5.63	7.83	24	1.96	2.61	1.16	7.85	0.63
Kundabhairepur	31.37	3.10	0.31	210	5.48	7.80	23	1.97	2.65	1.20	7.86	0.61
Madanpur Devrar	29.40	3.08	0.30	215	5.21	7.75	22	1.98	2.68	1.25	7.87	0.59
Ranjitpur	28.00	3.00	0.30	155	4.95	7.72	20	2.04	2.74	1.31	7.74	0.56
Katgharakhurd	25.94	2.92	0.29	210	4.72	7.76	19	2.10	2.82	1.37	7.73	0.55
Rampurustada	24.46	2.85	0.28	205	4.68	7.65	18	2.15	2.88	1.45	7.76	0.51
Ramnogar	21.61	2.78	0.27	208	4.43	7.62	17	2.27	2.92	1.58	7.69	0.48
Chaubepur	20.86	2.55	0.25	185	4.28	7.53	15	2.38	3.07	1.73	7.67	0.43

1	2	3	4	5	6	7	8	9	10	11	12	13
Amkol	17.70	2.42	0.24	190	4.07	7.51	14	2.45	3.12	1.81	7.65	0.38
Kunwarpur-I	15.83	2.35	0.23	195	3.95	7.50	13	2.56	3.18	1.95	7.55	0.37
Semari Dargar	11.49	2.27	0.22	190	3.82	7.45	13	2.68	3.26	2.08	7.52	0.32
Ballipur	8.33	2.15	0.21	192	3.67	7.46	12	2.78	3.55	2.17	7.47	0.28
Saifullaganj	5.76	1.92	0.19	175	3.52	7.32	11	2.85	3.45	2.38	7.43	0.25
Karondipur	3.77	1.79	0.17	185	3.45	7.21	10	2.96	3.60	2.42	7.36	0.19
Basauthi	1.08	1.54	0.15	195	2.78	7.13	9	3.10	3.78	2.65	7.32	0.10

TABLE 3.23 : **Correlation Matrix**

Sl. No.	Variables	Y	X_1	X_2	X_3	X_4	X_5
Y	% of the other types of Waste Lands	1.000					
X_1	Slope	0.990***	1.000				
X_2	Ruggedness Number	0.992***	0.997***	1.000			
X_3	Mean Annual Rainfall	0.097	0.144	0.092	1.000		
X_4	Drainage Density	0.976***	0.972***	0.978***	0.067	1.000	
X_5	Quality of Water in pH	—0.124	—0.127	—0.137	0.595**	—0.089	1.000
X_6	Behaviour of Water Table	0.970***	0.963***	0.971***	0.066	0.988***	—0.101
X_7	Nitrogen of the Soil	—0.996***	—0.995***	—0.994***	—0.127	—0.972***	0.138
X_8	Phosphate of the Soil	—0.991***	—0.998***	—0.998***	—0.116	—0.980***	0.124
X_9	Potash of the Soil	—0.996***	—0.977***	—0.996***	—0.123	—0.976***	0.129
X_{10}	pH Value of the Soil	0.988***	0.980***	0.980***	0.160	0.971***	—0.045
X_{11}	Land Concentration	0.993***	0.997***	0.998***	0.123	0.982***	—0.111

Sl. No.	Variables	X_6	X_7	X_8	X_9	X_{10}	X_{11}
Y	% of the other types of Waste Lands						
X_1	Slope						
X_2	Ruggedness Number						
X_3	Mean Annual Rainfall						
X_4	Drainage Density						
X_5	Quality of Water in pH						
X_6	Behaviour of Water Table	1.000					
X_7	Nitrogen of the Soil	—0.968***	1.000				
X_8	Phosphate of the Soil	—0.969***	0.994***	1.000			
X_9	Potash of the Soil	—0.970***	0.998***	0.995***	1.000		
X_{10}	pH Value of the Soil	0.966***	—0.985***	—0.985***	—0.984***	1.000	
X_{11}	Land Concentration	0.971***	—0.995***	—0.985***	—0.996***	0.982***	1.000

* Significant at 10% level of significance.
** Significant at 5% level of significance.
*** Significant at 1% level of significance.

and it is very low (7.13% to 7.32%) in Saifullaganj, Korondipur and Basauthi sample villages which is given in Table 3.22. When high quality of water in pH is used for irrigational purposes, it spreads salinity into the soils which is a hindrance to crop production. It can be seen from Table 3.23 that there is significant negative correlation between the quality of water in pH and other types of waste land. This shows that the above-discussed hypothesis is not statistically supported and this variable has the least importance in explaining the variation of other types of waste land.

Behaviour of water table is found very low in those villages where the percentage of other types of waste land is very high and it is very high in the villages having a very low percentage of other types of waste land because the areas having a high water table have more moisture in soil but in the low water table areas water percolates down due to the soft nature of the soil and such type of soils have low moisture and have low fertility status. It has been recorded from Table 3.23 that there is significant positive correlation at 1% level of significance between behaviour of water table and percentage of other types of waste land.

3.9.4 Pedogenic (Nitrogen X_7, Phosphate X_8, Potash X_9 and pH Value of the Soil X_{10})

The variables, nitrogen, phosphate and potash of the soil, are not considered for further analysis because these have been considered to estimate the fertility requirement of the soil.

The pH value of the soil is recorded very high (Table 3.22) and very low in those villages where percentage of other types of waste land is also very high and very low respectively because of the excessive amount of soluble salts which appear to be a hindrance to crop production. It is also proved from the correlation matrix (Table 3.23) that there is significant positive correlation between percentage of other types of waste land and pH value of the soil, which is significant at 1% level of significance.

3.9.5 Anthropogenic (Land Concentration/Gini's Co-efficient)

Land concentration is recorded very high (0.59 to 0.63) in villages with very high percentage of other types of waste land sample villages and it is low (0.10 to 0.25) in low percentage of other types of waste land sample villages (Table 3.22) because large size land holding farmers do not pay much attention to improve such land as they have other good quality of land where they are sure of certain output with limited input, even if they are unable to cultivate all the land. This is proved statistically (Table 3.23), that there is significant positive correlation between percentage of other types of waste land and land concentration significant at 1% level of significance.

3.9.6 Stepwise Regression Analysis

The system of explanation by endogenic variables discussed above shows a higher degree of multi-collinearity among themselves. To remove inconsistencies arising out of this problem of multi-collinearity stepwise regression analysis is attempted taking other types of waste land as independent variable (Y) and environmental and anthropogenic factors as independent variables (X). The result of the stepwise regression analysis gives an inter-correlation matrix which is given in Table 3.24. In three sets, stepwise regression analysis is attempted paying attention to those variables which have less multi-collinearity. It is found that at certain steps R^2 starts declining, thereafter the analysis gets truncated at these steps. The variables which are found on the optimal regression line in different sets of stepwise regression analysis are as follows.

In the first set of stepwise regression, the most dominant variable is X_{11} (land concentration) which explains 99.7% total variation of other types of waste land, while in the second step variable X_3 (mean annual rainfall) and in the fourth step variable X_{10} (pH value of the soil) are incorporated but these variables do not explain variation in other types of waste land, and in these steps R^2 does not increase, and these variables are subsumed in variable X_{11}.

The dominant variable, in the second set of stepwise regression, is variable X_{11} (land concentration) which alone

TABLE 3.24 : **Other Types of Waste Lands : Results of the Stepwise Regression Analysis**

Variables	Intercept	Regression coefficeint	S.E.	T	R^2	Increase in R^2	$\overline{R}$	$\overline{R}^2$	F
Step 1) (11	1.22409	3.117	0.072	43.020***	0.997	—	0.99651	0.99303	1850.683***
Step 2) (11	1.10831	3.108	0.073	42.319***	0.997	0.000	0.99648	0.99297	914.499***
) (3		0.001	0.001	0.917					
Step 3) (11	0.88804	3.075	0.415	7.411***	0.997	0.000	0.99619	0.99239	558.195***
) (3		0.001	0.001	0.842					
) (10		0.031	0.383	0.081					
Step 1) (11	0.11401	0.323	0.006	51.256***	0.998	—	0.99753	0.99506	2627.132***
Step 2) (11	0.13117	0.324	0.006	54.507***	0.998	0.000	0.99785	0.99570	1498.350***
) (3		—0.000	0.000	—1.678					
Step 3) (11	0.05426	0.313	0.033	9.354***	0.998	0.000	0.99770	0.99541	925.910***
) (3		—0.000	0.000	—1.653					
) (10		0.011	0.031	0.350					
Step 4) (11	0.06205	0.311	0.039	8.078***	0.998	0.000	0.99749	0.99408	631.926***
) (3		—0.000	0.000	—1.471					
) (10		0.010	0.035	0.279					
) (6		0.000	0.001	0.099					

Step 1) (5	141.55522	6.948	2.603	2.669**	0.595	—	0.59484	0.33395	7.122***
Step 2) (5	131.16033	7.196	2.649	2.716**	0.625	0.03	0.58588	0.34325	3.839***
) (11		20.692	24.512	0.844					
Step 3) (11	220.40788	6.627	2.488	3.065***	0.716	0.091	0.6533	0.43076	3.848***
) (5		218.027	121.210	1.799*					
) (4		—40.506	24.432	—1.658					
Step 4) (11	—4.79133	7.394	2.762	2.677**	0.718	0.002	0.61882	0.3294	2.656*
) (5		188.856	171.541	1.101					
) (4		—41.582	25.897	—1.606					
) (10		31.946	126.619	0.262					
Step 5) (11	26.37762	7.444	2.950	2.523**	0.718	0.000	0.56748	0.32203	1.917*
) (5		191.891	183.096	1.048					
) (4		—44.825	41.646	—1.076					
) (10		28.257	138.110	0.205					
) (6		0.583	5.653	0.103					

* Significant at 10% level of significance.
** Significant at 5% level of significance.
*** Significant at 1% level of significance.

explains 99.8% total variation of other types of waste land. But in the second, third and fourth steps variables X_3 (mean annual rainfall), X_{10} (pH value of the soil) and X_6 (behaviour of water table) are incorporated but these variables do not explain variation in other types of waste land because in these steps R^2 does not increase which is recorded constant.

In the third set of stepwise regression most dominant variable is X_5 (quality of water in pH) with explains 59.5% variation of other types of waste land. In the second step variable X_{11} (land concentration) with variable X_5 explains 62.5% variation of other types of waste land. In the third step variable X_4 (drainage density) is incorporated which jointly which variable X_5 and X_{11} explains 71.6% variation of other types of waste land. In the fourth step variable X_{10} (pH value of the soil) jointly with variables X_5, X_{11} and X_4 explains 71.8% total variation of other types of waste land. In the fifth step variable X_6 (behaviour of water table) is incorporated but it does not explain variation of other types of waste land because at this step R^2 is not found increasing.

Thus it can be concluded that the hypothesis related to the variables of environmental and anthropogenic factors with other types of waste land is supported by this empirical exercise. The high degree of multi-collinearity among the variables suggests that the effects of other independent variables on other types of waste land concentration, quality of water in pH, drainage density and pH value of the soil.

3.10 CONCLUSION

One can conclude that a combination of environmental and anthropogenic factors play an important role in the spatial variation of waste lands. In the first set morphometric, climatic, hydrologic and pedogenic are the important factors, while the second set is anthropogenic.

The analysis of the correlation matrix and stepwise regression analysis provides that slope, behaviour of water table, drainage density and distance of a canal from the field are the dominating variables in explaining the variation of waterlogged land.

Important variables in explaining the variations of usar

land are pH value of the soil, behaviour of water table, drainage density and thickness of the top soil.

The dominant variables in explaining variation of banjar land are land concentration, pH value of the soil, slope and quality of water in pH.

The important variables are land concentration, behaviour of water table and pH value of the soil in explaining total variation of old fallow land.

The main factors in explaining variation of fallow land are land concentration, drainage density and pH value of the soil.

The important factors influencing kankrili land are pH value of the soil, quality of water in pH, behaviour of water table and depth of base rock and drainage density.

The important variables in explaining variation of ravine land are drainage density, pH value of the soil, quality of water in pH, behaviour of water table and slope.

The dominant variables in explaining variation of other types of waste land are land concentration, quality of water in pH, drainage density and pH value of the soil.

4

GENESIS OF WASTE LANDS

4.1 INTRODUCTION

The word genesis means origin, creation, to come into being or denotes mode of formation (generation). The problem of waste lands is not a present experience but was there in the medieval period but the term then indicated little used common land, usually the light soil which failed to yield to the medieval cultivators. Assuming that every phenomena over the earth's surface has a history of development the same is being applied to the waste lands which, in terms of Davis, is a product of the fundamental sequential development of land forms. Davis propounded that this development is mainly dependent on underlying structural variations, the varying climatic differences and, above all, the time that is taken by a particular process to achieve its predominance. Henceforth the third factor becomes a parameter of measurement of intensity of development. This is relative in the sense that two regions of the identical nature may depict varying intensities of development due to the complex nature of interrelated structure and process. The measurement of intensity of waste lands is associated with the factors responsible for this development.

The existing waste lands in the study area are being analysed through chronological sequence with respect to what they are today. If the severity of the problem is high, one can assume that this has resulted through sequential development of certain processes on various types of litho-structures over a certain period of time. Thus, in this case study the sequential development of certain processes is organised for certain waste lands and their genesis has been explained.

Sultanpur is not very heterogeneous in terms of physiography, but the inception there of various types of waste lands is interesting. This has been analysed in the last chapter through various environmental and anthropogenic factors. Here with the help of the same factors their genesis is being attempted.

Here it is to be clarified that the percentage area of the waste land and intensity may not necessarily correspond with each other, because only a small patch can exhibit higher intensity than a larger area with not so well marked development.

4.2 THE GENESIS OF WATERLOGGED LAND

The waterlogging problem in the district which is observed in various villages provides sufficient clue that waterlogging occurs due to an increase in the water table caused by either more seepage or more water stagnation/accumulation in the low-lying areas due to overirrigation and seepage from canals. There is no outlet for the accumulated water as the affected areas are either too near to the canals or these are surrounded by road and railway lines which restrict the outflow of water.

Based on the above observations, to trace to its origin the problem of waterlogged land the same high, medium and low percentage villages—Malapur Jagdishpur, block and tahsil Kadipur; Baijipur, block Bhadainya, tahsil Sadar Sultanpur; and Haripur, block and tahsil Kadipur—are selected respectively for which the causative factors are analysed through different sequential stages of development, viz. initial, intermediate and severity stages of waterlogged land development.

(1) The initial stage is the one in which the development of waterlogged land either just begins or is to start with available suitable conditions. This can be termed as stage of *limited percolation.* Factors remaining the same, there is little or no percolation downward and the process of low leaching is inhibited. The surfacial processes have gained momentum mainly due to a small amount of sloped. The water table and drainage density is recorded low. The waterlogged area of

village Haripur is located far away from the canal which provides less possibility for the accumulation of excessive irrigated water through seepage from unlined canal banks and its drains. Thus in the selected village it is observed that all the processes discussed above give rise to the initial stage of waterlogged land. In this village, in the low-lying areas water gets accumulated through excessive irrigationing water and there is less chance of more accumulation of water as the water level increases it flows out towards the south-east side of the village but once this outflow is blocked the waterlogging problem may increase. In the initial stage it is noted that there is little percolation downward due to the high water table which is comparatively nearer to the surface.

(2) The subsequent stage may be termed as stage of *seepage* but the areas may be referred to as little more severe than the inception stage. But it is observed that this patch is surrounded marginally by such land as is not all that well developed in terms of severity and hence is categorised under inception stage. Thus for analysing the intermediate stage of waterlogged land, village Baijipur of block Bhadainya, tahsil Sadar Sultanpur is selected where inception and intermediate stage of waterlogged land development is recorded in the periphery and core of the affected area respectively. The development of such waterlogging is so recorded because there is little downward percolation, amount of the slope is recorded more than the inception stage, drainage density has also increased, while the affected area is nearer to the canal than in the previous stage. Thus all these factors in combination have provided suitable conditions for the accumulation of water while there is no outlat and the water table is also recorded high through surface and subsurface drainage. Due to the favourable factors, the water table is found to be increasing and the area of the village where waterlogging problem started a few years back is now considered to be at the intermediate stage and peripheral areas also seem to be suffering from the same problem and as such waterlogged land of the village has increased.[1]

1. The information is collected by personal communication with the villagers and increase in the area is not mentioned because the farmers were unable to give the proper recorded area, which is also not available from the Patwari's record at village level.

(3) This is the stage where the percentage of waterlogged land is higher than the other villages and this stage is termed as *surface accumulation* stage of waterlogged land. To analyse the severity of waterlogged land village Malapur Jagdishpur of block and tahsil Kadipur has been selected because this village has the highest percentage of waterlogged land than any other village of the district, where it is observed that all the processes pertaining to waterlogging have become more severe than in the previous stages. It is recorded that in this village the slope amount is very high, which provides better conditions for the overland flow down the slope and the drainage density is also high. The waterlogged patch is much nearer to the Sharda canal which actually passes through the village. The water table is recorded very high because of the near saturation of the soil and more surface and sub-surface drainage is low. All the drainage lines of the canal and its banks are unlined which provide excessive water for the increasing water table. Because the area is low-lying, water accumulation is more through excessive irrigational water and at the same time there is no outlet as it is surrounded by road and canal on the three sides. Thus in this village, in much of the peripheral area waterlogged land is in the initial stage and after that intermediate stage and in the middle of the affected area severity of the problem is observed. Thus the factors as observed are involved in excessive surface accumulation resulting in waterlogging.

4.3 GENESIS OF USAR LAND

Those soils which have an excessive amount of soluble salts and exchangeable sodium are known as usar. It is recorded in most of the areas that where the precipitation is less than the evaporation in association with morphometric factors it leads to impeded drainage and the ground water hydrology characterised with high water table enriched in soluble salts in some areas is mostly responsible for the accumulation of salts through transportation and deposition in other places thereby causing manifestations of saline characters in the soil. In other areas it is also observed that prolonged use of saline irrigation water also brings in the development of soil salinity. To analyse the genesis of usar land, village Ghazipur (block Bhadar, tahsil Amethi), village Tewaripur (block Dhanpatganj, tahsil Sadar Sultanpur) and village Baithu (block Kurebhar, tahsil Sadar Sultanpur) are selected. They have high (39.96), medium (17.85) and low (0.74) percentage of the usar land respectively.

(1) The nature of the usar soil is very much associated with pH value of the soil. In the initial stage it is observed, in village Baithu, that there is high drainage density and slope which provide less suitability for the accumulation of salt which gets washed off along with the soil. The pH value of the soil and rise in the water table is low but the thickness of the top soil is high. This subsequent slow rise in the water table exposes a few salts and it gives rise to a few salt encrustations over the earth surface. In the initial stage in which residue is associated with *surface wash* morphometric factors lead to impeded drainage and hydrologic characteristics associated with high water table and low quality of water in pH are mainly responsible for the less accumulation of salts. The deposition of salts in this area is low, while transportation is recorded more due to which pH value of the soil is low.

(2) In the intermediate stage, thickness of the top soil lessens and the water table is nearer to the surface. Village Tewaripur of block Dhanpatganj, tahsil Sadar Sultanpur is selected to analyse the usar land development where slope is recorded lower than in the initial usar land village. Drainage density is low and water table has increased. The value of the quality of underground water in pH has increased and the thickness of the top soil is lower than in the previous stage. Thus in this stage all the factors which are discussed above are more favourable in increasing the salts. This may be termed as the stage of *salt encrustation.*

(3) It has already been discussed that the pH value of the soil is very much associated with the usar land. In the severity stage, where the salt concentration process is high, village Ghazipur of block Bhadar, tahsil Amethi is selected for analysing the usar land development. This area has high pH value of the soil, where precipitation is usually less than

evaporation in association with certain morphometric variables (low slope is helpful in accumulation of salts) leading to impeded drainage and the hydrologic factor (drainage density is low, quality of water in pH is high and rise in the water table is high) gives rise to much salt encrustation over the earth surface. Thickness of the top soil is very low. Thus in this stage pH value of the soil is very high which is a hindrance to the crop production. This stage is the *usarisation stage.*

4.4 GENESIS OF FALLOW, OLD FALLOW AND BANJAR LANDS

The fallow land is that type of land which is left uncultivated for 1—2 years, old fallow land is that land which is left out of cultivation for 2—5 years and banjar land is that land which is left out of cultivation for more than 5 years due to poverty of the cultivators, large size of land holdings, mismanagement and carelessness of the cultivators. Natural factors are also observed to be the dominating factors, i.e. inadequate supply of water, climate and, in a few areas, silting of canals. The main problem which is recorded in many areas is the deteriorating soil conditions of the land due to any reason of either overcroping or mismanagement of the land It is recorded that the formation of fallow land is mainly due to the deteriorating soil conditions or inability of the soil to produce the crop because of which the farmers have left it out of cultivation and when it is left for more than 5 years the problem becomes severe. So it can be considered fallow land as the initial stage (one year), old fallow land as intermediate stage and banjar land (more than 5 years) as the severe stage of waste land development. Thus the same patch of land after remaining uncultivated for a certain period of time is classified as fallow, old fallow and banjar while the processes in the transformation from the initial stage to the severe stage are the same. To analyse the genesis of fallow, old fallow and banjar land village Bhadainya (block Bhadainya, tahsil Sadar Sultanpur), village Raibigo (block and tahsil Kadipur) and village Piparpur (block Bhadar, tahsil Amethi) are selected respectively because these villages have the highest percentage of the respective waste land types to the total area of the respective villages.

(1) The genesis or beginning of the fallow is considered from one year and is left uncultivated for two years. Most of the villages suffer from much weed infestation and deteriorating soil conditions and that is why the present stage according to the occurrence is termed as structural deterioration of the soil. The farmers leave the land uncultivated for maintaining the soil fertility status. In case the farmers are unable to bring the current fallow under cultivation for two years, such land is categorised as fallow land, considered as the initial stage of waste land development. It is recorded from village Bhadainya that the behaviour pattern of morphometric (slope is low, drainage density is high, rise in the water table is high, quality of water in pH and pH value of the soil is low but the concentration of land, i.e. Gini's coefficient is high) and anthropogenic factors are very much responsible for the development of fallow land. Keeping all the factors aside it is observed that there is deterioration in the fertility status of the soil, where loose top soils are washed off and there is much weed infestation. It is found that farmers are least bothered to maintain the fertility status of the soil due to the large size of land holdings.

(2) The intermediate stage is of the old fallow land which is left out of cultivation for 2 to 5 years. For analysing the genesis of old fallow land village Raibigo is selected where much soil wash is recorded. The old fallow land is the subsequent stage of the fallow land. It is recorded that morphometric, hydrologic, pedogenic and anthropogenic, i.e. land concentration factors are the same as in the case of fallow land. In this stage land is left out of cultivation for more than 2 years but less than 5 years due to the same deteriorating conditions of the fertility status of the soil, where loose top soil is washed off through water and wind and land is left uncultivated due to the poverty of the farmers, carelessness of the cultivators, large size of land holdings and inadequate water supply.

(3) The severe stage (according to the time period) is the banjar land which is left out of cultivation for more than 5 years and is termed as *barrenness stage*. The banjar land is subsequent and final stage of old fallow land. At this stage the

problem has become more severe because the land is left out of cultivation for more than 5 years. To analyse the severe stage of banjar land development village Piparpur is selected where the morphometric factor (rise in the pH and high pH value of the soil and high land concentration) provide enough clue that the soil is oversupplied with the soluble salts because of which the fertility status of the soil has gone down and the land is left out of cultivation for more than 5 years. The farmers who own large size land holdings do not pay much attention to maintaining the fertility status of the soil. Even due to mismanagement, overcropping and inadequate water supply, the land is left out of cultivation for more than 5 years and the problem worsens due to continuous degradation of the top structure.

4.5 GENESIS OF KANKRILI LAND

The kankrili land is that type of land where a high percentage of calcium carbonate nodules called kankars are found. This has an excessive amount of soluble sodium in the exchange complex and is dominated by the salts of the carbonates and bi-carbonates of mainly sodium where the pH value of the soil is found to be very high. To analyse the various stages of kankrili land development, village Sarwan (block Kurebhar, tahsil Sadar Sultanpur), village Samsapur (block and tahsil Kadipur) and village Seurchamurkha (block Kurebhar, tahsil Sadar Sultanpur) are selected for inception, intermediate and severity stages respectively having low, medium and high percentage of kankrili land to the total area of the village. It is already discussed that growth in overall area may be taken as the index for the development of kankrili land. Regarding their intensity higher percentage is equated with severe development.

(1) The initial stage of the kankrili land, as observed in village Sarwan, had no kankar nodules earlier but have been brought through erosion from nearby elevated areas and this stage is termed as *slope wash stage*. The elevated areas, according to the collected information, are the old built houses but now are

ruined and between these elevated kankar lands, the lowlying area is occupied by a tank. From these elevated areas, kankar nodules are washed off by both wind and water and are deposited in adjoining areas layer, upon layer, year after year. Due to the new kankar deposition there is impermeable strata near to the surface. Due to the behaviour of morphometric factors (slope and ruggedness is low, drainage density, quality of water in pH is low, rise in the behaviour of water table is high) and due to the new deposition of kankar nodules the depth of base rock is found low. The soil is not much enriched in soluble salts and that is why the pH value of the soil is low.

(2) The intermediate stage is the stage of *stratification of calcium carbonate nodules* where subsequent erosion and sedimentation of kankars from the elevated areas leads to comenting of the kankar. The deposition of kankar nodules strata is recorded more which is impermeable and carbonate nodule formation through carbonation is more due to early deposition. The slope is recorded high, drainage density is high, quality of water in pH is high and rise in the behaviour of the water table is low than the initial stage of kankrili land development. Due to thicker strata of calcium carbonate nodules depth of the base rock is high. The pH value of the soil is high because the soil is enriched with soluble salts while the presence of calcium carbonate nodules is high.

(3) In the severe stage of kankrili land *intensive calcification of kankar nodules* is observed. In village Seurchamurkha the top kankar nodules are wahsed off from the elevated areas and are deposited in nearby lowlying areas for the past more than 15 years, because of which sedimentation or kankar nodules layer by layer is more, depth of base rock is high, slope and drainage density is high, rise in the behaviour of water table is very low than the intermediate stage. The quality of water in pH and pH value of the soil are high because there is an excessive amount of exchangeable soluble sodium, and due to much sedimentation of kankar nodules impermeable strata is very near to the surface. The soil has low water intake rate and hydraulic conductivity resulting in poor moisture transmission characteristics. The soil below the kankar pan is sandy clay and gradually the percentage of clay diminished with more formation of kankar nodules. There is deficiency of available calcium ,

nitrogen and micronutrients, while there is nutritional disorder. The presence of calcium carbonate nodules is recorded at a depth of more than 5 feet. The soluble salt content in the soil is higher, which accumulates in the sub-soil through the process of capillary action which is much helpful in the formation of carbonate nodules through the carbonation process.

4.6 GENESIS OF RAVINE LAND

The ravines often designated as bad lands are found all along river Gomati. There is substantial area in Sultanpur along the Gomati which is turned into the ravine land and this problem is acute in the district. Most of the land of the district had been so badly damaged by erosion and cut by ravines that agriculture of any type is impossible. The genesis of ravine land is analysed through various stages of ravine land development where Chandpur Sadar (low percentage ravine land village), block Kurebhar, tahsil Sadar Sultanpur; villages Semari Dargar (medium percentage) and Ratanpur (high percentage of ravine land) of block Kurebhar and tahsil Sadar Sultanpur are selected for analysing initial, intermediate and severe stages of the ravine land development respectively.

(1) The initial stage, which is *rain wash stage*, of ravine land development is analysed in village Chandpur Sadar which is 8 km. away from river Gomati. Due to the surface wash, the loose sands are washed off from a higher elevation downward forming narrow hills. The slope in the village is recorded 7.12%, drainage density 2.41%, behaviour of water table 10 feet while the pH value of the soil is 7.23. In the initial stage the width of the narrow gullies is less than 1.5 metres while the depth of the gullies is less than 3 metres.

(2) The intermediate stage is of the *entrenching of channels* and this stage is analysed in village Semari Dargar. In this stage slope is found to be 8.61%, drainage density 4.38, behaviour of water table has gone down, pH value of the soil is 7.46. It is recorded that in this stage due to severe erosion and sheet wash depth of the gullies increases from 3 metres to 6 metres and the width of the gullies also increases from 1.5 metres to 3 metres and in this stage more entrenching of channels is recorded.

(3) In the severity stage *severe bank erosion* is observed. Due to the severe erosion banks have collapsed, which is observed in village Ratanpur. The slope is observed to be very high, 9.39°, drainage density 5.20, which is responsible for maximum erosion, the behaviour of water table is very low, pH value of the soil is found to be 7.67. It is recorded in the sample village that most of the gullies have a depth of more than 6 metres and width of more than 3 metres because of severe bank erosion and this area is easily affected by the floods in river Gomati. In this stage severe lateral and bank erosion is recorded due to which the depth and width of the gullies have increased much more than the intermediate stage of ravine land development.

4.7 GENESIS OF OTHER TYPES OF WASTE LAND

The other types of waste land does not include waterlogged, usar, banjar, old fallow, fallow, kankrili and ravine lands as discussed above. The other type of waste land is mainly observed near the roads, canals, railway lines and also those areas which are away from the ravines of the Gomati but temporarily left out of cultivation. It is recorded in many areas that due to surface wash the loosened sands are washed off from higher elevation downward forming small narrow nalas. Due to much erosion even the good soils are eroded because the soil is very soft, due to which fertility status of the soil has gone down. Slope is 3.12°, drainage density is high (5.63), low water table is (24 feet) and same (7.8) quality of water in pH and pH value of the soil. These are all observed in village Dadara where the percentage of other types of waste land is higher then in other villages to the total area of the village. Associated with the abovementioned environmental factors, the land is left out of cultivation due to the large size of land holdings, poverty of the farmers, mismanagement and carelessness of the farmers.

It is to be confessed here that a proper developmental process could not be achieved for this particular category of waste lands.

4.8 CONCLUSION

Development through sequential stages is attempted for all the waste lands associating them with causative factors and processes. Waterlogging develops through water table increase and the usar through increasing salinisation, the fallows through deteriorating soil structure, ravines through severe bank cutting and the kankrili through calcification.

5

WASTE LANDS CAPABILITY CLASSIFICATION

5.1 INTRODUTION

The waste land and their determinants and genesis as they have been identified and discussed in the preceding pages necessitate that the problem be explored still further to assess their potentialities in terms of agricultural productivity, so that necessary measures may be suggested to improve their condition. Once the factors and the processes of their formation are identified, it becomes easier to predict their present capability.

The development of waste lands operationalised through an interaction mechanism between nature, man and technology proceeds through a triangle forces. Here nature indicates the direction of its optimal appropriation and lays down the expanding limits of human freedom. The technology aids natural processes, bends them to human purposes and this expands the range of freedom for agricultural activity. The institutions are conducive to or restrictive of man-nature interactions. It is through this system of inter-dependence that the formation of waste lands begins. Thus the development of waste lands is essentially a function of differential doses of technological inputs interacting with enviromental constraints of varying severity under the inhibiting influence of institutional factors of different intensities. Where all these three operate negatively i.e. where institutions are restrictive, environment impermissive and the level of technology is low, in such conditions waste land increases. It is recorded that interaction of man with his technology is very limited in Sultanpur District while the severity of nature is not much but the farmers are still following the traditional type of agricultural activity where technological innovations are very limited and are ignorant about such innovations for interacting with the waste. Thus it can be said that limitations of waste lands depend upon the severity of environmental constraints or how successfully a farmer with his low level of technology is able to tackle such constraints.

The physical environment of man consists of land forms along with other natural resources. The land forms provide man space, food and vast variety of raw materials which are used for satisfaction of human needs. Thus among such land forms waste lands which are temporarily out of cultivation play a strategic role in the determination of man's economic progress through increasing agricultural productivity. It is imperative that under the severity of environment, associated with man's ambitious utilisation of the land the increase in waste lands can be immense. Natural erosion is a continuing process and will go on into the future regardless of anything man may do. Erosion that we are so seriously concerned with is an abnormal and undesirable process started by man's objectives and is subject to his control. Unchecked erosion may produce poverty and undermine the strength of the nations. Hence conservation should be the main aim. So it is essential to find out capability classification of different waste land types in waste land reclamation planning and development for optimal agricultural use and adequate solutions.

To make a sound and conservation plan, one should establish the land capability which has to do with the suitability of land for specified uses. The land capability classification is a systematic arrangement of different kinds of land according to those properties that determine the ability of land to produce on a virtually permanent basis. Suitability for cultivation is assumed to include the use of machinery—at least of plough, tillage implements and harvesting equipment and the capacity for at least a moderate yield of one or more crops with suitable treatment and protective measures. The restrictions imposed by natural land characteristics necessarily affect (a) the number and complexity of the corrective practices to be used, (b) the productivity of the land and (c) the intensity and manner of land use. The classification is made and used for practical purposes, which is the selection and application of land uses and treatments that will, while using the land, keep it in condition for long-time production. The latter will involve control of erosion, conservation of rainfall and maintenance. The classification is arranged according to a number of categories. The first category covers the land suited for cultivation and not for cultivation, the second category covers 8 land capability classes. The classes under this category are differentiated according to the degree of limitation in land use, imposed by the land itself. The eight land capability classes range from the best and most easily formed land (class 1) to land which has no value for cultivation, grazing or forestry but may be suitable for wildlife, recreation and water shed protection.[2] Thus the capability is the inherent quality of waste lands to perform at a given level of general agricultural use while suitability is a statement of adaptability of a given area for a specific land use type.[3] On the other hand waste land classification is a method of grouping waste lands on the basis of their classes assessed in relation to their suitability to agricultural uses envisaged in the plan. It relates to the morphometric, climatic, pedogenic and hydrologic and other similar environmental conditions affecting the agricultural use and productivity.

The waste lands classification will help to organise significant factors for conservation. It is an imperative grouping and grading of soils, according to their potentialities and limitations, their capacity of producing agricultural crops and responsiveness to management practices. This is a system to make a sound and complex form conservation plan to find out the suitability of waste lands for specific areas and its agricultural uses. Thus the methods of waste lands capability classification is a systematic segregation of different kinds of waste lands which are distinguished from one another by variation in the kind and degree of use imposed by soil characteristics morphometric, hydrologic, climatic and other environmental

2. "A Manual on Conservation of Soil and Water", *Handbook of Professional Agricultural Workers*, U.S.D.A., Oxford Publishing Co., 1984, pp. 23-28.
3. F.A.O., 1976, pp. 64.
(A Framework for land evaluation, *Soil Bulletin* No. 1).

factors. Thus the fundamental purpose of waste land capability classification is to utilise the waste land resources according to their capabilities through rational planning.

The waste land capability classification in Sultanpur District is mainly based on the environmental parameters of morphometric (slope and ruggedness number) climatic (rainfall), hydrological (drainage density, quality of water in pH and behaviour of water table), inherent soil charateristics and personal field observations. On the basis of this the following five capability classes have been ascribed to the various waste lands in the district:

Capability Class I—Very easily reclamable
Capability Class II—Easily reclamable
Capability Class III—Reclamable with certain difficulty
Capability Class IV—Reclamable with moderate difficulty
Capability Class V—Reclamable with great difficulty

5.2 Very Easily Reclamable

Very easily reclamable waste land types with a few limitations are those types of waste lands where the environmental severity is very low and man with the low level of technology is able to

Waste Lands Capability Classification

Capability Class I	Capability Class II	Capability Class III	Capability Class IV	Capability Class V
Very easily reclamable	Easily Reclamable	Reclamable with certain difficulty	Reclamable with moderate difficulty	Reclamable with great difficulty
Fallow land and other types of waste land	Old Fallow land	Water logged Banjar land	Usar Land and Revine land	Kankrili land

interact for availing waste land resources for agricultural use. In this category, other types of waste land and fallow land (1—2 years) are identified because these two waste land types experiencc less severe natural constraints and there was a time when these waste land types were under agricultural use but due to deteriorating in soil conditions are presently left out of cultivation.

The other types of waste land is very easily reclaimable because the morphometric variables slope and ruggedness number in the selected villages, are very low and erosion by surface drainage is very low which does not lead to too much loss in lowering the fertility status of the soil. Rainfall and other climatic factors are suitable for agricultural use but the land needs a strong bunding to check the erosion by rainy water. Drainage density is low which will cause less erosion, water table is low but for the irrigational purposes water can be easily lifted. The soil is sufficient in the organic matter. There is little calcium in the soil which contains little amount of sand while lower horizons are coarse in texture and the pH value of the soil is normal which can be easily maintained for agricultural use using organic and inorganic amendments. The quality of the soil is very good and it is suitable for agricultural use.

The fallow land can be very easily reclamed because the slope and ruggedness number are low, climate is suitable, drainage density is low and behaviour of the water table is suitable for agricultural use but the land is left out of cultivation due to low fertility status of the soil and large size of land holdings. These types of land can be very easily reclaimed through high doses of organic and inorganic amendments. The quality of soil is good and suitable for agricultural purposes and such parctices on these waste land types will be fruitful because these lands are left out of cultivation only for 2 years due to low fertility status of the soil caused by mismanagement and carelessness of the farmers.

5.3 Easily Reclaimable

In this category old fallow land is categorised because this waste land type is lefe out of cultivation for more than

2 years but less than 5 years due to low fertility status of the soil and can be easily reclaimed because the morphometric charaeteristics are suitable, drainage density is low which leads to less erosion. The water table is low but through tube-well the irrigational facilities may be managed. pH value and low fertility status of the soil can be improved through organic and inorganic amendments because the quality of soil is otherwise good and suitable for agricultural practices but due to overcropping, large size of land holdings, mismanagement and carelessness of the farmers the fertility of the soil has gone down which can be easily reclaimed through proper attention and technology doses. The old fallow land is recommended for agricultural uses and agricultural practices will be fruitful in these waste lands.

5.4 RECLAMABLE WITH CERTAIN DIFFICULTY

In this category those waste land types are classified where environment is severe to a certain extent and man with proper utilization of technology may interact with such waste land types to avail these resources. These types of waste lands may be reclaimed with certain difficulty. Banjar land and waterlogged land are categorised in this capability class of waste lands. These waste land types were used by the farmers but certain environmental calamities, mismanagement and carelessness of the farmers have created the problem in agricultural practices by which the farmers were bound to leave the land uncultivated.

The banjar land may be reclaimed with certain difficulty because of the very low fertility status of the soil. The morphometric, hydrologic and climatic factors are not so problematic in improving the banjar land. In the pedogenic factor pH value of the soil is high but can be managed with certain difficulty through organic and inorganic amendments. Firstly, these lands were kept as current fallow for maintaining the fertility status of the soil but may be due to poverty or carelessness of the farmer or large size of land holdings, farmers were not bothered to bring such lands under cultivation and in succession they are left out of cultivation for more than 5 years. That is why such lands can be brought under cultivation only with certain difficulty, because for 5 years these have remained uncultivated. Reclamation may be little problematic.

The waterlogged land can also be brought under cultivation with certain difficulty because there is no outlet. More drainage lines (both artificial and natural) and rise in the water table through surface and sub-surface drainage both has caused water accumulation through continuous flow from the upper areas to the low-lying areas. This may be the rain water or excessive irrigational water through canals and seepage from the unlined canal banks. This can be managed with certain difficulty because it needs total overhauling of the drain for providing an outlet to the accumulated water. The soil characteristics can be managed through organic and inorganic amendments. This problem can be solved by constructing drains for the outflow of the water in which government officials of the district, as well as the cooperation of all the villagers is needed because the waterlogging problem is not personal problem of the farmers but is a problem for all the villagers. So consciousness among all the villagers can be brought through proper training and suggestions for the proper water utilisation.

5.5 RECLAIMABLE WITH MODERATE DIFFICULTY

The waste lands which are reclaimable with moderate difficulty are those types of waste lands where nature is severe than the reclaimable with certain difficulty waste land type and man is less permissible with his low level of technology to avail the waste land resources. The usar land and ravine land can be reclaimed with moderate difficulty. It is recorded from the sample villages that in these two waste land types severity of environmental constraints is high and man is less permissible to interact with his low level of technology to avail the waste land resources for agricultural purposes.

The usar land can be brought under cultivation with moderate difficulty. The morphometric variables (slope and ruggedness number) are recorded low which provide suitable conditions for a rising trend in the water and accumulation of salty water from other areas. Increase in the salt is high through capillary action, due to which pH value of the soil is very high and fertility of the soil has gone down. Rainfall also helps in accumulation of salts in the low-lying areas. There are a few drainage lines which do not help in transportation of salt but provide drainage towards the low-lying areas which helps in deposition of salts. High quality of water in pH is harmful for the crop and it cannot be used for irrigational purposes, so the irrigational facilities have to be managed with difficulty from other sources through cemented minor drains. The rise in the water table, enriched in soluble salts, is often responsible for the accumulation of salts, which badly affects the fertility status of the soil. The usar land has an excessive amount of salts and exchangeable sodium, i e. high pH value of the soil which is a hindrance to crop production. There is deficiency of available calcium, nitrogen and micro-nutrients. There is nutritional disorder and it has an extremenly low water intake and hydraulic conductivity resulting in poor moisture transmission characteristics. There is also presence of calcium carbonate nodules which are totally unfit for cultivation. Thus all the abovementioned morphometric, hydrologic and pedogenic factors are severe, including the quality of soil which is not fit for agricultural practices but, with moderate difficulty, adopting high level of technological inputs with the help of soil and agricultural scientists, usar land can be reclaimed.

The ravine land can also be reclaimed with moderate difficulty. In the ravine land slope and ruggedness number are high leading to more erosion resulting in an increase in the depth and width of the ravines. Rainfall is helpful in erosion and much rainfall through erosion causes more formation of ravines and gullies and even the open sky with high wind velocity is helpful in the erosion because in the affected areas soils have more content of sand and such soils have low moisture retaining capacity which is a hindrance to crop production. The sub-soil water table is found to be very low which is unfit for crop production but can be improved with moderate difficulty with proper irrigational facilities. There are sandy and sandy loamy soils in the ravine land. There are numerous sandy tracts which provide much evidence of soil erosion. Thus the abovediscussed factors are highly responsible in the formation of ravine land and are a hindrance to agricultural activities. Such land can be reclaimed with moderate difficulty with a high level of technology and with proper knowledge of recommended agricultural practices according to the suitability and quality of the soil with other environmental constraints which are severe.

5.6 RECLAIMABLE WITH GREAT DIFFICULTY

Reclaimable with great difficulty are those waste lands where environmental factors are very severe, institutions restrictive or man with low level of technology is not able to exploit these waste land resources. Kankrili land is classified in this category and it can be reclaimed with great difficulty. The morphometric constraints (slope and ruggedness number) are high enough not to permit man agricultural activity and with his low level of technology hard kankar pan cannot be easily levelled. The rainfall is helpful in eroding the kankar nodules from elevated areas and depositing these in surrounding areas but again it helps calcification of kankar nodules; through which kankar strata spreads above and below the surface. The high drainage density is responsible for erosion, transportation and deposition of kankars in the surrounding areas. High quality of water in pH is much helpful in the fromation of calcium carbonate nodules because this water has an excessive amount of soluble salts. The water table is low because the soil has low water intake rate and hydraulic conductivity. The soil has poor physical conditions in the kankrili land. There is exchangeable and soluble sodium with high pH value of the soil. Hence the pedogenic characteristics of the kankrili land are very much severe. Thus it has been recorded that environmental severity in the kankrili land is high and it can be reclaimed only great difficulty by, firstly, separating kankars from the land and, secondly, with very high doses of technological inputs with proper management of agricultural practices.

5.7 CONCLUSION

Thus the waste land capability classification is the best illustration of the usefull work which is done in Sultanpur District. It may be remarked that this type of waste lands classification is of immense importance and of practical utility in the waste land management and agricultural development.

Although it needs enormous labour and technical knowledge together with huge funds, it would help greatly in predicting the suitability and adaptability of waste lands to various agricultural purposes. On the basis of capability classification of waste lands, it has been recorded that fallow land land other types of waste land can be very easily reclaimed, old fallow land can be easily reclaimed, banjar land and waterlogged and can be reclaimed with certain difficulty, while usar land and ravine land can be reclaimed with moderate difficulty and kankrili land can be reclaimed with great difficulty.

6

REMEDIES FOR RECLAMATION OF WASTE LANDS

6.1 INTRODUCTION

As has been noted, the soil deterioration due to erosion ultimately leading to waste lands is one of the most pressing problems of the nation. Already it has ruined a good acreage of cultivable land and if unchecked will continue to affect the remaining land also. In short, unless we adequately safeguard our area of good land we may eventually face a serious shortage. Hence, the concept of conservation to protect the existing cultivable land and the reclamation of already depleted waste lands list in priority of the planning by the Government bodies.

The reclamation measures which have been adopted in different areas of Sultanpur District have been practised for ages, not on the scientific basis covering all types of waste lands, but only a few types, viz , usar, waterlogged and ravine. Some of the methods were successful and later on proved to be technically sound, whereas a few methods proved unsuccessful after some time because the reverted back to its original condition due to mismanaged and unscientifically adopted reclamation measures and lastly a few other measures failed to show any response from the land so treated. Thus the methods which are adopted for the reclamation measures should take into account the causes which led to the development of waste lands or deteriorating soil conditions of the land in association with morphometric, climatic, hydrologic parameters, including suitability and quality of the soil for producing a particular crop. This should not be taken to mean that such ameliorative methods are not worth any trial. In fact they have shown a remarkable progress under certain conditions as recorded from different types of waste lands during the reclamation period. But the input-out put will have to be balanced with benefits accrued from the land over a number of years. Thus in any scheme of reclamation, the time and involved expenditure for the reclamation are the additional factors to be considered. It has been recorded that the cheaper amendments are slower in reaction and produce uneconomic returns at the beginning. Again certain ameliorating factors may be available as surplus farm commodity or industrial waste in an area but may prove prohibitive or may be available with difficulty at another place. These factors have to be fully considered before appreciating any technique or methods of reclamation of any waste land types. Thus in practice reclamation measures for such deteriorated lands have to be properly judged in the light of the above basic conditions with special attention paid to environmental factors in association with anthropogenic factors which are responsible for the development of different waste land types.

The remedies for the reclamation of different waste land types in totality cannot be uniform because for different waste land types, different reclamation methods are required. In this chapter, the measures to be adopted to rejuvenate the 8 categories of waste lands are discussed. For the purpose 8 villages have been selected (of each waste land type) on the basis of the hightest percentage of waste lands among all the villages of the district from which the affected one-acre land has been taken for the adoption of reclamation measures to see practically how these methods yield good results from the reclaimed land. The study has been carried over for both the main cropping seasons, kharif and rabi; over one acre of land of the selected village. The various reclamation remedies adopted at the time of reclamation for different types of waste land are elaborated below.

6.2 REMEDIES FOR RECLAMATION OF WATERLOGGED LAND

The remedies for the reclamation of waterlogged land have been experimented on one acre of land, in village Malapur

Jagdishpur (block and tahsil Kadipur), having the highest percentage of waterlogged land to the total area of the village, which is shown in Map 16. The essential components of the waterlogged land reclamation which have been adopted step by step at the time of reclamation are as follows:

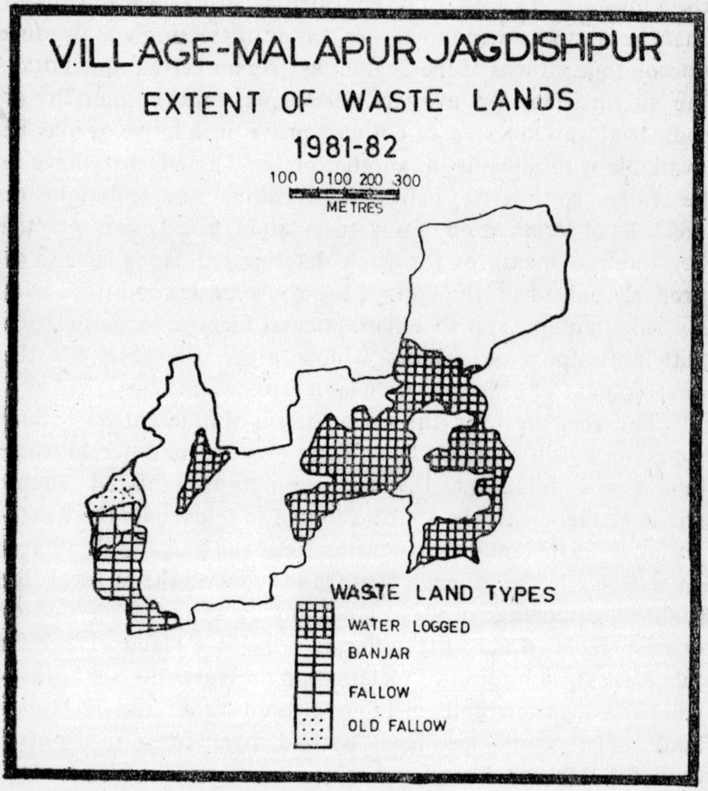

MAP 16

(1) Village Malapur Jagdishpur is surrounded on two sides by the road and on one side by the canal. Due to its being low-lying, excessive irrigation and rain water from all the sides gets accumulated as there is no outlet for the excessive water to drain out. The Government had constructed a drain for the outflow of the accumulated water but as it was not properly

planned, the problem remained the same. The new measure which is adopted in the village is to redig the same drain and widen and deepen it and join it with another nala away from the village through which excessive accumulated water can easily flow out.

(2) The levelling of the land is done, providing strong bunds along the field and the farmer is trained for the proper implementation of the correct doses of fertilizer irrigation and other technological inputs. At 15 cms below from the level of the field, in accordance with the slope, small drains are made nearer to the field for the outflow of the excessive water.

(3) After testing of the soil, and according to the requirement of the soil, to produce paddy crop and keshari crop, nitrogen, phosphate and potash fertilizers are applied.

(4) Unlined canal banks are not cemented on all sides but in few areas *culabas* and *gul* of the canal drains are properly made pucca near the affected area to check the seepage of water.

(5) Regular meetings are held between the personnel of the canal irrigation department and the farmers for the use of canal water, in irrigation, in a proper quantum according to the requirement of the field.

(6) Paddy is grown as a first crop during the kharif season selecting high yielding tolerant variety and timely plantation of relatively older seedlings.

(7) In the second rabi season keshari is preferred mainly because in this field, paddy takes more time and more moisture of the soil, while keshari crop does not need much water and is found to be ready within the limited time. This crop rotation is to be followed for at least three years and the farmers are advised not to leave the land fallow for a long time.

6.3 REMEDIES FOR RECLAMATION OF USAR LAND

Various reclamation measures for the usar land are practised in village Ghazipur (block Bhadar, tahsil-Amethi) having the highest percentage (depicted in Map 17) of the usar land to the total area of the village.

At the Central Soil Salinity Research Institute (CSSRI),

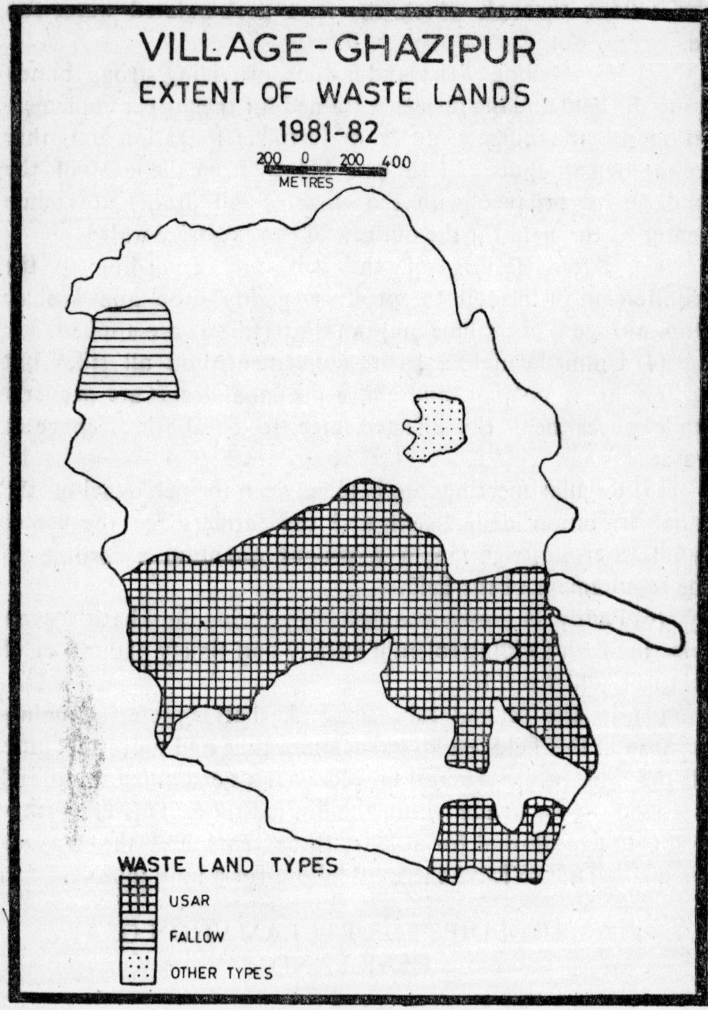
VILLAGE-GHAZIPUR
EXTENT OF WASTE LANDS
1981-82
200 0 200 400
METRES
WASTE LAND TYPES
USAR
FALLOW
OTHER TYPES

MAP 17

Karnal, in 1969. scientists have developed a package of technology for the reclamation of saline soils. Measures for reclaiming usar land in India have been several including (i) Mechanical, (ii) Leaching, (iii) Application of chemical amendments, (iv) Growing of salt tolerant crops.

Adoption of suitable methods have to be encouraged, taking into consideration such factors as total cost, availability of good quality of water, type of soil, nature of terrain, local cropping practices, etc. The earlier workers like Leather (1983, 1914) and Voslcker (1893) and others as mentioned in the literature survey (Chapter I) had tried preparation of drainage, leaching, application of heavy doses of organic matter which have also been found feasible in Uttar Pradesh and Punjab by many workers. However reclamation measures through application of organic matter alone is considered feasible only when the deterioration in the soil is found and beneficial effects appear to be of short duration.

Reclamation measures for the usar land are adopted practically in village Ghazipur, adopting all the practices of technology rather than to use the pick and choose approach. Availability of good quality water for irrigation is an essential prerequisite. The essential components of usar land reclamation which are adopted at the time of reclamation are as follows:

(1) Proper levelling, providing bunds along the field is done to obtain a uniform spread of water. The field is divided into convenient size plots with small earthen embankments in accordance with slope. Small suitable drains are made for the removal of excess rain and irrigation water. Leaching with rain water during the rainy season is carried over.

(2) Testing of the soil is done to know the nature of the soil and fertilizer requirement to produce crop. In the first season 40 quintals of gypsum is applied and alongwith this 50 kg. urea, 25 kg. Shakti Khad and 25 kg. of M. Potash and also compost is used. Lastly top dressing of fertilizer is done applying 25 kg. of urea. In the rabi season, for producing wheat crop a total of 75 kg. of urea, 50 kg. of M. Potash and 50 kg. of Phosphate fertilizer are applied. Easily decomposable organic matter (Dhaincha) is also applied before rain. Gypsum in powdered form is mixed with the upper 10 cms. of soil. After

application of gypsum, water is needed to stand for 10—15 days before paddy plantation. Pyrite can also be applied.

(3) In the first kharif season paddy is grown and the transplanting and growing of paddy crop for a few years is done until a good crop is obtained. An additional 25% nitrogen is added to the normal recommended dose without zinc application. The green manuring is followed by rabi crop.

(4) In the rabi season, wheat crop is grown applying 25% more nitrogen. High yielding tolerant variety of wheat crop is sown.

(5) Light but frequent irrigations is managed in rabi season, according to the requirement of the crop. It is much useful to grow dhaincha as a green manure crop when sufficient irrigation water is available. This crop rotation is to be followed for at least 3 years continuously and farmers are advised not to leave the land fallow for a long period.

The techniques discussed above could be suitably modified to meet the requirements of the soil. But before any reclamation project is taken into hand, there should be adequate levelling of the fields, improvement of the drainage and supply of the ample irrigation water of good quality. A project on the reclamation of such type of uncultivated land, big or small, can be an economic proposition if planned on scientific lines. In many cases deterioration caused to the soil by salt is more easily repairable than caused due to erosion. This is due to the fact that the land prices have gone up and there is scarcity of land, so the reclamation projects should be carried out to meet the requirement of the farmers.

6.4 REMEDIES FOR RECLAMATION OF BANJAR LAND

For the successful cultivation of the banjar land through chemical and technological amendments, an understanding of the behaviour of water table and knowledge of the quality of the soil are necessary. Various reclamation measures which are applied to reclaim one acre of banjar land in village Piparpur (block Bhadar, tahsil Amethi) (Map 18) are as follows:

(1) The field is equally levelled for ensuring uniform application of irrigation water and fertilizer, uniform crop growth and better reclaiming action of the amendment. The field is

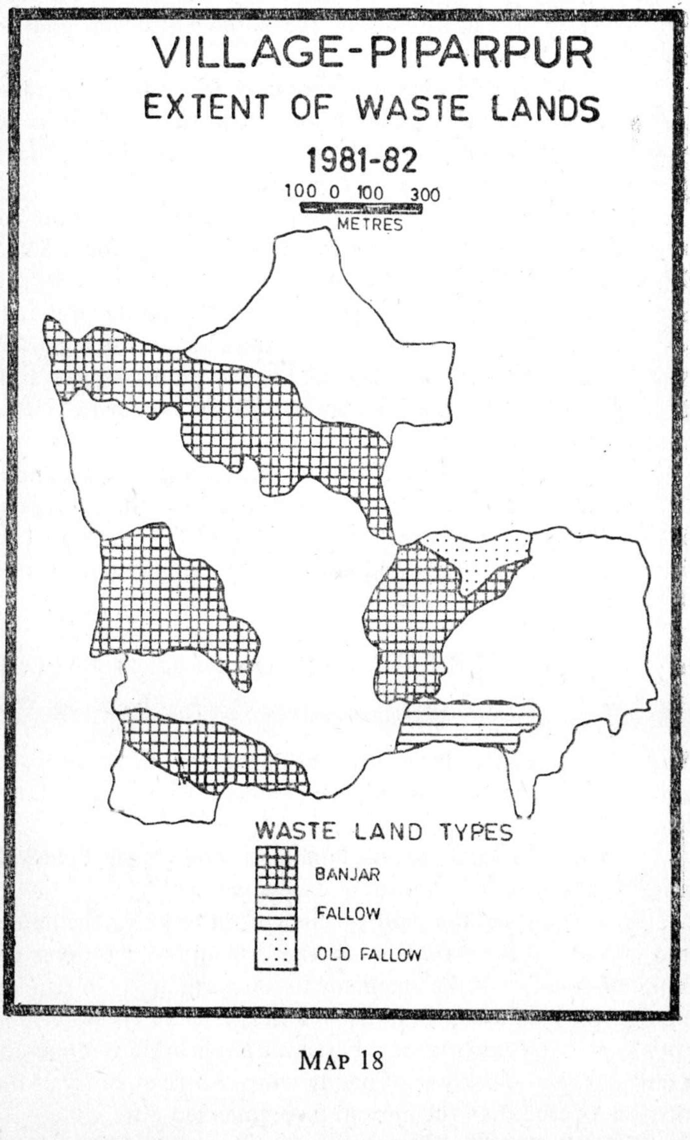
VILLAGE-PIPARPUR
EXTENT OF WASTE LANDS
1981-82
100 0 100 300
METRES
WASTE LAND TYPES
BANJAR
FALLOW
OLD FALLOW

MAP 18

divided into different size plots for judging proper levelling of the land, with proper bunds. The bunds are made because it is required essentially to prevent the entry of run-off water from outside area and also to check the outflow of water from the field.

(2) The surface drainage is necessary and it is made along the outside of the bunds, for providing proper outflow for the water. Good quality water is necessary for irrigation and it is being tested before its use in irrigation.

(3) Soil testing is done, and according to the nature and requirement of the soil to produce a good crop, urea, Shakti Khad and M. Potash is applied in both kharif and rabi seasons. Gypsum/pyrite is used and it is mixed in the upper 10 cm with light ploughing 15 to 20 days before sowing of the paddy crop. After the use of amendments excessive irrigation is done till 5-7 cm level is obtained. Green manuring is done after cultivation.

(4) In the first kharif season, paddy crop is sown and in the second rabi season wheat crop is sown of the high yielding tolerent variety. After reclamation, farmers are advised to follow this crop rotation and not to leave the land fallow for a long time.

6.5 REMEDIES FOR RECLAMATION OF OLD FALLOW LAND

Various measures have been adopted for reclaiming one acre of old fallow land in village Raibigo (block and tahsil Kadipur) (Map 19).

(1) Levelling and strong bunding is done in the field with suitable drainage for removal of excess water.

(2) Testing of the soil is carried out to know the nature and extent of the problem of the soil and to determine the required quantity of the amendments. According to the requirement of the soil, urea, compost and Shakti Khad are applied in both kharif and rabi seasons. Fertilizer application is done for a few days in the case of paddy crop. More quantity of the nitrogen is used than the normal recommended dose.

(3) In the first kharif season paddy crop and in the second rabi season, wheat crop is sown. The seeds in both the cases

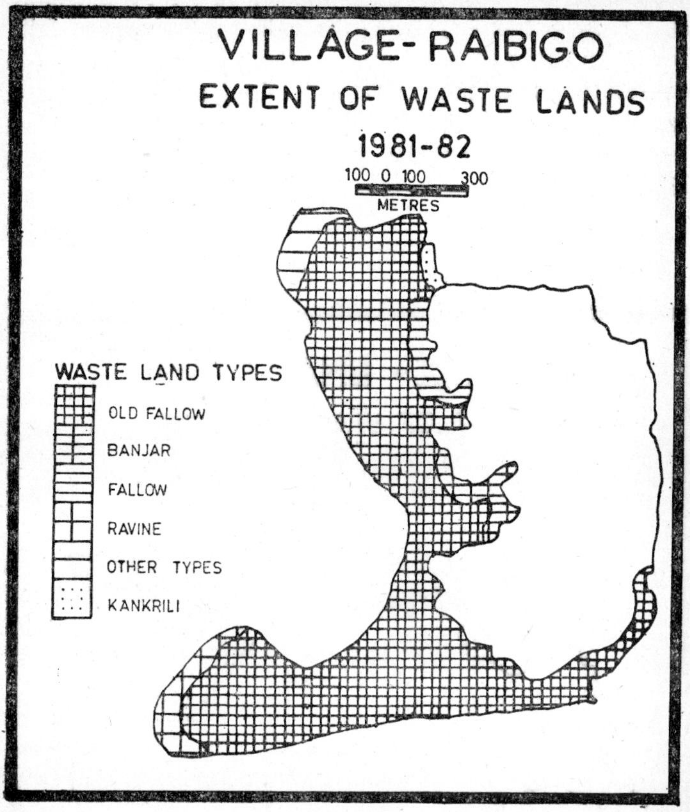

MAP 19

are of the high yielding varieties. According to the requirement of the crops irrigation facilities to the crops are provided, while green manuring is not done due to the scarcity of water. Lastly farmers are advised to follow this crop rotation and not to leave the land fallow for a long time.

6.6 REMEDIES FOR RECLAMATION OF FALLOW LAND

Various reclamation measures have been adopted to reclaim

one acre fallow land in village Bhadainya (block Bhadainya, tahsil Sadar Sultanpur) having the highest percentage of fallow land (Map 20) to the total area of the village. For achieving best results, all practices of the technology are adopted rather than to use the pick and choose approach. The adopted essential components of the fallow land reclamation are as follows:

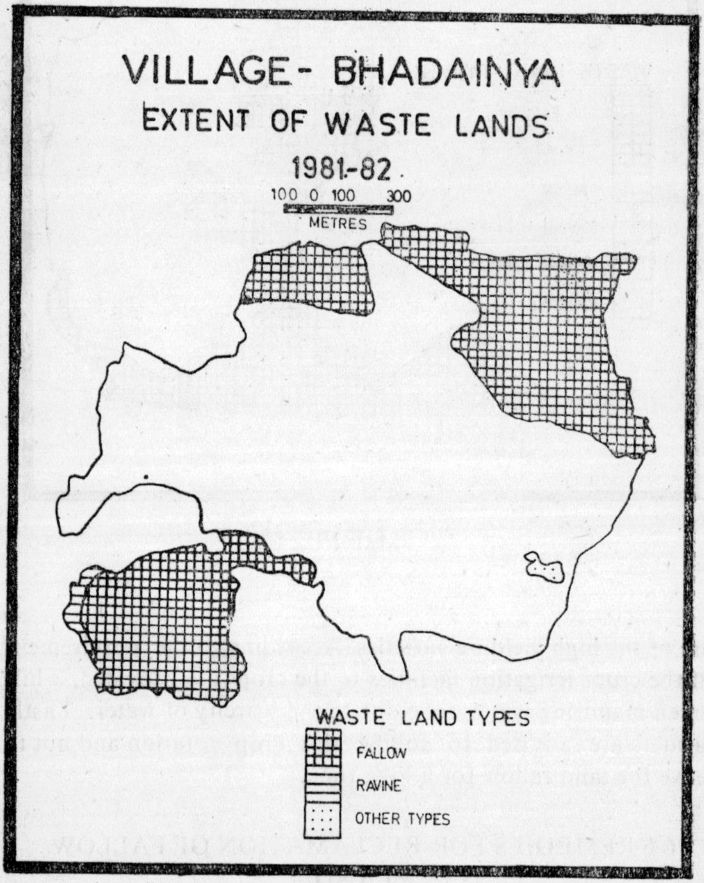

MAP 20

(1) Proper levelling of the land is done in order to obtain a uniform spread of water and fertilizer. The field is divided into different size plots with small earthen embarkments and is provided suitable drainage for the removal of excess water.

(2) According to the results of the tested soil and to the known nature of the soil, urea, Shakti Khad, potash and compost amendments are applied in the kharif and rabi seasons.

(3) As the first crop paddy is sown during the kharif season, using more nitrogen, while the farmer is advised to sow the paddy crop for a few years until a good crop is obtained. In the rabi season, high yielding variety of wheat crop is sown followed by green manuring. Light but frequent irrigation is provided to the wheat crop according to its requirements.

(4) Again after the rabi season Dhaincha as a green manure crop is grown and testing of the soil is done to know the nature of the soil. Lastly the farmers are advised to follow this crop rotation at least for three years and not to leave the land fallow for a long time.

6.7 REMEDIES FOR RECLAMATION OF KANKRILI LAND

Kankrili land reclamation is practised in one acre of land in village Seurchamurkha (block Kurebhar, tahsil Sadar Sultanpur (Map 21). The measures which are adopted at the time of reclamation are as follows:

(1) The compact clay or kankar pan in the soil is processed by deep ploughing to break the pan and after this kankars are thrown out of the field. Proper levelling of the land is done for equal distribution of fertilizer and irrigation water. According to the slope, land is divided into convenient size plots, with earthen embarkments. Proper strong bunds are also provided along the field. Suitable drains are made for the removal of excess water. Testing of the soil is carried out to know the nature and extent of the problem of the soil. According to the requirement of the soil urea and compost are applied for producing *chari* crop in the kharif season and barley crop in the rabi season. Early decomposable organic matter before the on-

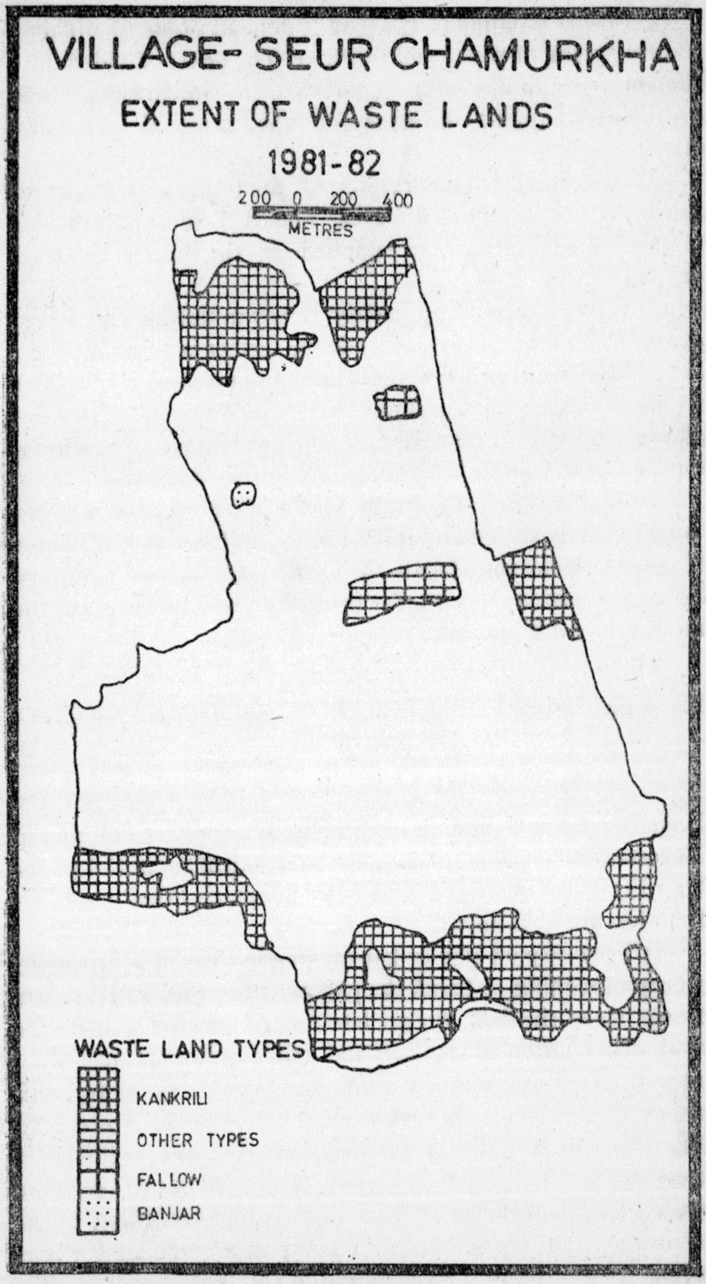
VILLAGE- SEUR CHAMURKHA
EXTENT OF WASTE LANDS
1981-82
200 0 200 400
METRES
WASTE LAND TYPES
KANKRILI
OTHER TYPES
FALLOW
BANJAR

MAP 21

set of the rains is applied. Farmers are suggested to grow fodder crops for a few years until a good crop is obtained because there is an uncertainty of a good outout from paddy and other crops, seeing the nature of the land.

(2) Thus in the first kharif season fodder crop, *chari*, is sown and in the second rabi season, barley crop is sown. Light but frequent watering is provided to the barley crop according to its requirement.

(3) The farmers are advised to follow this crop rotation at least for 3 years and not to leave the land uncultivated for a long time.

6.8 REMEDIES FOR RECLAMATION OF RAVINE LAND

It is important to know the extent of the land under ravines and a survey should be conducted to assess whether the ravine in question is suitable for agricultural purposes after reclamation or is to be devoted for forests or grass lands. During the survey, information about the depth and width of the ravines and the type of soil should be obtained. It would be desirable that an aerial survey of the contiguous ravines is conducted and topographical sheets be prepared to know the nature of the gullies and ravines. However with proper reclamation measures which may be expensive such land may become suitable for agricultural purposes. Anyhow for the practical exercise various measures of reclamation are carried out with the help of the Kamla Nehru Farm Science Centre, Sultanpur, in village Ratanpur (block Kurebhar, tahsil Sadar Sultanpur) in one acre of ravine land (Map 22) because this village has the highest percentage of ravine land to the total area of the village. The following reclamation measures are adopted to reclaim one acre of ravine land:

(1) Mechanical (tractor) reclamation is used for levelling the ravine land and bullock power is also used for terracing and levelling of the land. According to the slope and terraces, the field is divided into different size plots and strong bunds are constructed around the field to check the erosion and water flow. Suitable drains are made for removing excess water. Land preparation and ploughing is done according to the

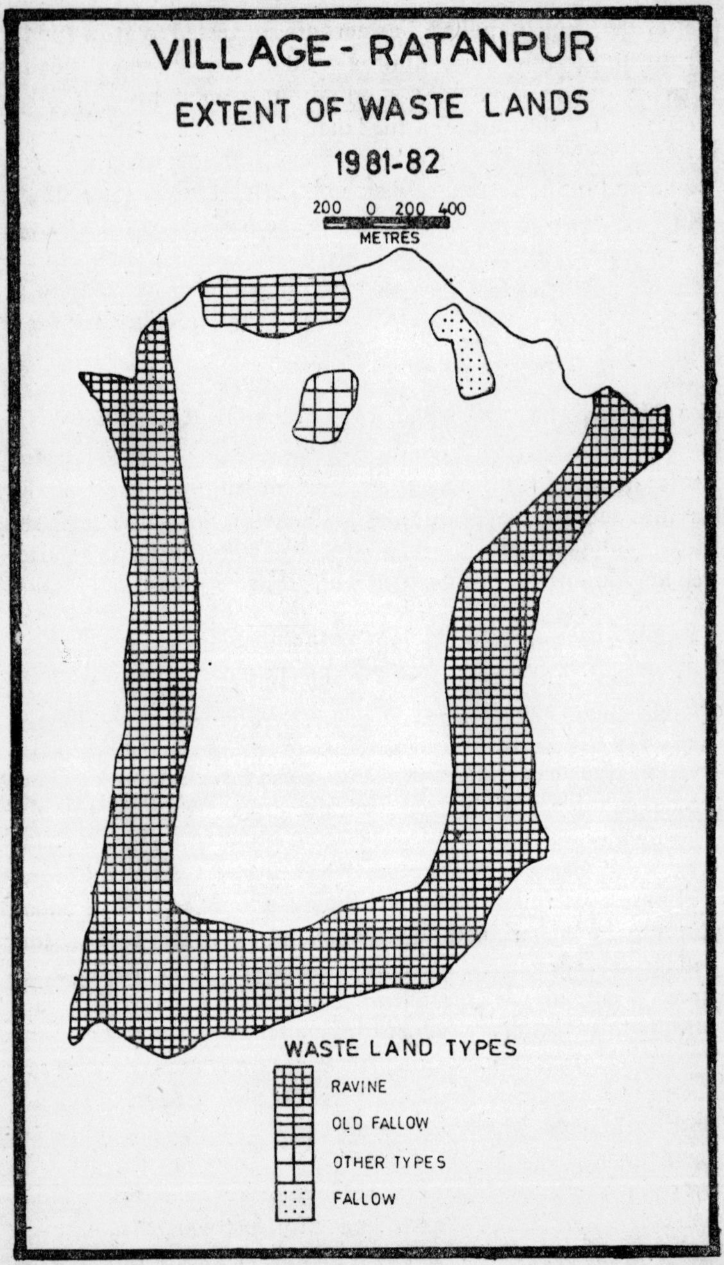
VILLAGE - RATANPUR
EXTENT OF WASTE LANDS
1981-82
200 0 200 400
METRES
WASTE LAND TYPES
RAVINE
OLD FALLOW
OTHER TYPES
FALLOW

MAP 22

nature of slope. Diversion channels are constructed to reduce the erosion through gullies. A dam was constructed earlier to check the erosion and flooding.

(2) Testing of the soil is done to know the requirement of the soil to produce a particular crop and according to its requirement urea, Shakti Khad, potash and compost are applied in both khariff and rabi seasons.

(3) In the first khariff season maize fodder crop is sown and in the second rabi season *barseem* fodder crop is sown providing light but frequent irrigation according to the requirement of the crop because this soil does not have much moisture retaining capacity. In the summer season, after the rabi crop *Dhaincha* as a green manure crop is sown. The farmers are advised to follow this crop rotation for at least 3 years and not to leave the land fallow for a long time.

(4) The other simple measure for the ravine land reclamation is to stop grazing there and provide suitable outlets for 'excessive run-off'. Controlling of soil erosion may itself be sufficient to reclaim such land and measures discussed above may be adopted for the reclamation of ravine land.

6.9 REMEDIES FOR RECLAMATION OF OTHER TYPES OF WASTE LANDS

The reclamation measures have been practised in one acre of other types of waste lands in village Dadara (block and tahsil Musafirkhana) (Map 23) because this village has the highest percentage of other types of waste land to the total area of the village. Adopted reclamation measures for other types of waste land are as follows:

(1) Proper levelling of the land is done to ensure equal distribution of water, organic/inorganic amendments and strong bunding is done along the field. The field is divided into different size plots with earthen embarkments and suitable drains are provided for the removal of excess water.

(2) Testing of the soil is done to know the nature of the soil and, according to the requirement of the soil to produce a particular crop, urea, Shakti Khad, potash and compost are applied in the kharif and rabi seasons.

(3) In the first kharif season, jowar crop and in the second rabi season *barseem* fodder crop are sown. According to the

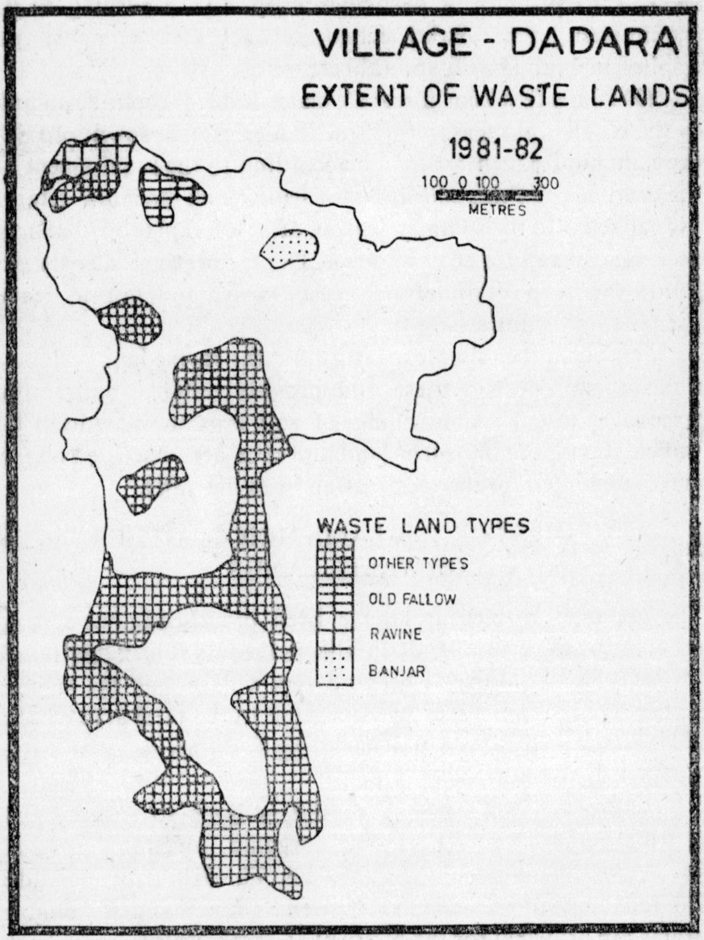
VILLAGE - DADARA
EXTENT OF WASTE LANDS
1981-82
100 0 100 300
METRES
WASTE LAND TYPES
OTHER TYPES
OLD FALLOW
RAVINE
BANJAR

MAP 23

requirements of the crop light but frequent irrigation is provided.

(4) The farmers are advised to follow this crop rotation for at least 3 years and not to leave the land fallow for a long period. If the erosion is more, it may be checked by constructing a dam, growing grasses, stopping of grazing and providing outlets for the excessive run-offs. Controlling of the soil erosion may itself be sufficient in reclamation of such types of waste land.

6.10 CONCLUSION

The above discussed reclamation measures can be suitably modified to meet the requirements of the soil, and waste land resources can be made available for agricultural purposes. But before any reclamation plan, it should be checked that there is adequate levelling of the field, improvement of the drainage and supply of ample irrigation water of good quality. A project on reclamation of any waste land type, big or small, can be an economic proposition if planned on scientific lines. The fact that land prices have gone up and there is scarcity of cultivable land more such reclamation projects should be undertaken with the help of government officials. Reclamation of any type of waste land is a physical problem. The general procedure followed to reclaim these lands is very simple. Thus, where the reclamation is not possible within the means available, suitable management practices will have to be developed, capable of coexisting with any type of waste land problem for some sort of economically worthwhile agriculture, through such methods as proper selection of crops, special planting and irrigational techniques. A list of such practices is given here: (a) Land preparation and tillage. (b) Seed-bed preparation and planting, (c) Irrigation, leaching and drainage, (d) Crop rotation, (e) Continuous cropping, (f) Application of amendments, (g) Afforestation and pasture development.

Apart from the reclamation methods, certain preventive measures and management practices should also be adopted. There are other important methods as intensive pre-irrigation, soil surveys and mapping of sub-soil characteristics, in regard to permeability, and to indicate areas where the water table is expected to rise. The extent of rejuvenation for gully formation can be explored from topo-sheets and aerial photographs, so that these factors may be taken into account while designing the irrigation system, such as lining sections of the channels, where the soil is excessively porous and the sub-soil is impermeable. Thus, if all these measures are adopted, they could lead to the extension of the cultivated area of Sultanpur District.

7

ECONOMICS OF WASTE LAND RECLAMATION

7.1 INTRODUCTION

An attempt to reclaim the waste lands in Sultanpur District can be described as essential and worthwhile from the point of view of the development of national wealth through increasing agricultural production by increasing the cultivated area. Table 2.2 shows the acreage of various types of waste lands distribution in Sultanpur District. It has been observed that improvement facilities and technical research have now reduced the necessity of allowing lands to be left uncultivated. Several technological practices may be used to maintain the deteriorated fertility status of the soil which is discussed in the preceding chapter.

Though the pressure of population is increasing at an increasing rate the extension of cultivation on uncultivated land is stagnant for several years ago. This is due to the fact that exploitation of such uncultivated lands has not been within the reach of the ordinary farmer. It is also a fact that such land is inferior in quality and is unsuitable for agricultural purposes. The reclamation of any waste land type is time-consuming and costly for a farmer and any reclamation which accentuates soil erosion cannot be desirable from the long-term point of view. On the other hand reclamation of waste lands is beneficial to farmers, who otherwise do not accentuate erosion and in fact enhance agricultural potential of a region. For a better adoption of the remedial measures by the farmers, the pros and cons should be made amply clear to them. The expenditure involved in practising such remedial measures, though costly in the initial stages, later on, may become an asset to the farmers. Such a diffusion of ideas to the farmers is possible once the economics of waste lands is achieved. Hence the only charm in reclaiming such lands is to achieve are additional agricultural production which is very necessary for our country.

7.2 ECONOMICS OF WASTE LANDS RECLAMATION

The various sets of waste lands discussed earlier require different types of remedial measures for their reclamation, and it is necessary to know the cost of reclamation. The economics of waste land reclamation for different waste land types poses different problems of management and reclamation measures. So the cost of reclaiming different waste land types also varies.

An attempt has been made to explain the average expense involved in the reclamation of one acre (0.4 hectare) of different waste land types, viz. waterlogged land (village Malapur Jagdishpur), usar land (village Ghazipur), banjar land (village Piparpur), old fallow land (village Raibigo), fallow land (village Bhadainya), kankrili land (village Seurchamurkha), ravine land (village Ratanpur) and other types of waste land (village Dadara) of Sultanpur District with the help of the land owner in association with government officials of the district. The input-output analysis has been made after reclamation of each waste land type. The main factors influencing the economics of waste land reclamation are the cost of reclaiming these lands and anticipated benefits on account of reclamation in terms of additional agricultural production. Investigation on the economic produces for reclamation of different waste land types has been carried out over a peroid of year i.e. kharif and rabi seasons, with the help of government officials of Sultanpur District.

7.2.1 Waterlogged Land

The reclamation of waterlogged land is a very costly affair and it requires total overhaul of the drainage system. It cannot be achieved by individual means and efforts. This can be achieved only by large-scale planning at government level. Thus the Government's intervention in reclaiming such waste land types is very necessary.

Village Malapur Jagdishpur in Kadipur Tahsil has been selected for analysing remedies and economics of waterlogged land reclamation. The abovementioned village is surrounded on two sides by the road and on one side by the Sharada Sahayak canal aligned along the inclined slope. In this village, there is a low-lying area where excessive water (rain and irrigation) accumulates and it has no outlet. Due to this, most of the area of the village is affected by it. It has been recorded from field observations and collected information from the villagers and Government officials that there is a drain for flushing out the accumulated water to save the area from the waterlogging problem. But the problem has remained the same because the drain was not properly made and due to excessive water collected through rain and irrigation and seepage, the water table is recorded very high. The new measure for solving this problem is to re-dig the same drain through widening and deepening and increasing its length according to the topographic conditions of the area. The drain dug out should be joined to another nala away from the village with a proper draining system. The involved expenditure in re-digging such a drain is approximately Rs. 30,000. By this it is recorded that the waterlogging problem of the village will be solved but it would entail as a permanent expenditure of Rs 1,000 per acre. This problem can be solved by large-scale planning at the government level and collective efforts of all the villagers.

Table 7.1 (A) shows that the preparation of a drain for the outlet of the accumulated water costs Rs. 1,000 per acre of land which is a fixed expenditure. The other expenditures are for land preparation, fertilizer application, transplanting of seedlings including nursery cost, harvesting and threshing which cost Rs. 1,325 and including the permanent expenditure of the drain preparation it costs Rs. 2,325. This is the expenditure in the first season (kharif) of reclamation cycle from which 14 quintals of grain and 16 quintals of straw are produced which bring in Rs. 1,820 for grain and Rs. 160 for straw respectively. Thus the total output achieved is Rs. 1,980 [Table 7.1 (B)]. In the first season of reclamation cycle there is a profit of Rs. 655

TABLE 7.1 (A): **Expenditure Involved in Reclaiming Waterlogged Land (one acre), Village Malapur Jagdishpur, Tahsil Kadipur. First Reclamation Cycle—Kharif season (Crop, Paddy)**

Operation	Human Labour (in mandays)	Wages Paid Rs.	Draught* Employed and payment made Rs.	Material used		Total Expenditure in Rs.
				Name	Cost	
Drain Preparation (Permanent Expenditure)						1000
Land Preparation	2	14	36			50
Fertilizer Application	3	21		Fertilizers (N;P;K.)	420	441
Transplanting of Seedling including nursery cost	20	140		Paddy Seedling	500	640
Plant Protection	2	14		ZNSo and Blit Ox	40	54
Harvesting	12	84				84
Threshing	8	56				56
Total Input	Permanent and current expenditure					=2325
	Current expenditure only					=1325

**Draught* : Shows hire charges for a pair of bullocks and plough paid @ Rs. 2 per hour.

TABLE 7.1 (B) : **Output per Acre in Kharif Season from the Reclaimed Waterlogged land, Village Malapur Jagdishpur**

Crop	Yield (in qtl.)		Prevailing Prices (in Rs. per qtls.)		Output (in Rs.)		Total	Profit from	Loss including Per-
	Grain	Straw	Grain	Straw	Grain	Straw	Output (in Rs.)	Current Expenditure (in Rs.)	menent and Current Experditure (in Rs.)
Paddy	14	16	130	10	1820	160	1980	655	345

excluding the permanent expenditure of drain preparation; when this cost is added there is a loss of Rs. 345.

In the second reclamation cycle (rabi season) keshari is sown without the initial fixed expenditure. It can be seen from Table 7.2 (A) that there is an expenditure of Rs. 465 which includes drain maintenance, land preparation, seedling, harvesting and threshing. The total output achieved from this crop is worth Rs. 1,540 which is given in Table 7.2 (B) in which the cost of grain (Rs. 1,500) and straw (Rs. 40) are included. Thus there is a profit of Rs. 1,075 to the farmer. It is learnt from the above exercise that the reclamation of waterlogged land is profitable from the second season onwards and various projects must be undertaken to reclaim such type of land in which farmers can benefit a lot and agricultural productivity can also be increased simultaneously. Though the initial expenditure in terms of total overhauling of the drainage is more in the long-term it may prove profitable in terms of agricultural production.

7.2.2 Usar Land

It has already been discussed that in usar land reclamation operations like levelling, bunding, land preparation and the use of gypsum and pyrite as well leaching and flushing are processed. The application of fertilizer differs in quantity and quality according to the requirement of the soil, hence the cost of reclamation also varies. According to the usarization extent, the prevailing cost of commodities and labour have been assessed in reclaiming one acre of usar land of village Ghazipur (tahsil Amethi) in which an effort has been made to chart out the input-output balance in both the kharif and rabi seasons.

Usar soils with reh encrustations have a high soluble salt content where white encrustations and powder appear on the soil surface in the summer season. Such type of land can be reclaimed with good quality of water alone by leaching, flushing and improving the drainage system. It is given in Table 7.3 (A) that the cost of reclaiming and sowing of paddy crop is Rs. 1,468 per acre of usar land with predominantly high sodium content and above 8.5 pH value of the soil. These lands

TABLE 7.2 (A) : **Expenditure involved in Reclaiming Waterlogged Land (one acre) during second Season of Reclamation Cycle, Rabi Season (crop, Keshari) Village Malapur Jagdishpur**

Operation	Human Labour (in mandays)	Wages Paid Rs.	Draught* Employed and payment made Rs.	Material used		Total input Rs.
				Name	Cost Rs.	
Drain Maintenance	2	14				50
Land Preparation	2	14	24			38
Seedling	2	14		Keshari seed	260	274
Harvesting	8	56				56
Threshing	5	35	12			47
Total Expenditure						465

**Draught* : Shows hire charges for a pair of bullocks and plough paid @ Rs. 2 per hour.

TABLE 7.2 (B) : **Output per acre in Rabi Season (Second Reclamation Cycle) from the Reclaimed Waterlogged Land Village Malapur Jagdishpur**

Crop	Yield (in qutls)		Prevailing Prices (in Rs. per qutls)		Output (in Rs.)		Total output Rs.	Profit Rs.
	Grain	Straw	Grain	Straw	Grain	Straw		
Keshari	15	4	100	10	1500	40	1540	1075

TABLE 7.3 (A) : **Expenditure involved during the First Season of Reclamation Cycle (one acre), Kharif Season (Crop, Paddy), Village Ghazipur, Tahsil Amethi Usar Land**

Operation	Human Labour (in mandays)	Wages paid Rs.	Draught* Employed and payment made Rs.	Material used: Name	Material used: Quantity	Material used: Cost Rs.	Total Expenditure
Levelling and Bunding	7	56	20				76
Land Preparation	4	32	60				92
Gypsum Application	5	40		Gypsum	40 Q.	200	240
Impounding of water	2	16		Irrigation		44	60
Fertilizer Application	3	24		Urea	50 kg.	124	
				Shakti Khad	25 kg.	64	
				M. Potash	25 kg.	30	
				Compost		50	292
Transplanting of Seedling including nursery cost	15	120		Paddy Seedling		250	370
Cultural Practices	1	8					8
Top Dressing of Fertilizer	1	8		Urea	25 kg.	62	70
Plant Protection	1	8		Zn So and Blitox		42	50
Irrigation	2	16		Tubewell	5 Hrs.	50	66
Harvesting	8	64					64
Threshing	10	80					80
Total Expenditure							=1468

**Draught* : Shows hire charges for a pair of bullocks and plough paid @ Rs. 2 per hour.

are characterised by black encrustations appearing on dry soil surface. The abovementioned expenditure in reclaiming usar land in the first kharif season includes levelling and bunding, land preparation, gypsum application, impounding of water, fertilizer application, transplanting of seedling including nursery cost, cultural practices, top dressings of fertilizer, plant protection, irrigation, harvesting and threshing. It has been recorded from Table 7.3 (B) that the total money output achieved in the first (kharif) season from the paddy crop is Rs. 1,995 which is made up Rs. 1,875 from grain and Rs. 120 from straw. Thus in the first kharif season from the paddy crop a farmer gets a profit of Rs. 527 on an investment of Rs. 1,468 on the reclamation of an acre of usar land.

In the second (rabi) season of reclamation cycle, in the same field wheat crop is sown and the total expenditure involved in reclaiming and other added inputs in sowing the wheat crop is Rs. 1,260, which covers land preparation, seedling cost, fertilizer application, cultural practices, top dressing of fertilizer, irrigation, harvesting and threshing, as given in Table 7.4 (A) and it can be seen from Table 7.4 (B) that the total output of Rs. 2,360 from sowing wheat crop which includes Rs. 2,080 for the grain and Rs. 280 for the straw according to the prevailing prices. It is observed from the above table that a farmer gets a profit of Rs. 1,100 in the second (rabi) season of reclaimed usar land where wheat crop is sown.

Thus the above instance seems to be very favourable and the return may be expected to exceed the investment in the successive season. It can be concluded that any attempt in reclaiming usar land is sure to be successful in terms of input-output balance in the long run. Thus the value of production will cover the cost of reclamation in a short period.

7.2.3 Banjar Land

The economics of banjar land reclamation is worked out at one acre of land of village Piparpur (tahsil Amethi) and it has been recorded from Table 7.5 (A) that there is a total expenditure of Rs. 1,214 in reclaiming one acre of banjar land in the first (kharif) season of the reclamation cycle, sowing paddy crop. The total expenditure includes levelling and bunding, land pre-

TABLE 7.3 (B) : **Output per acre in Kharif Season from the Reclaimed Usar Land**

Crop	Yield (in qutls)		Prevailing Prices (in Rs. per qutl.)		Output (in Rs.)		Total Output Rs.	Profit Rs.
	Grain	Straw	Grain	Straw	Grain	Straw		
Paddy	15	12	125	10	1875	120	1995	527

TABLE 7.4(A) : **Expenditure involved during the Second Season of Reclamation Cycle (one acre), Rabi Season (Crop, Wheat), Village Ghazipur.**

Usar Land

Operation	Human Labour (in man-days)	Wages Paid Rs.	Draught* Employed and payment made Rs.	Material used			Total Expen-diture Rs.
				Name	Qty.	Cost Rs.	
Land Preparation	4	32	60				92
Seedling Cost	2	16		Wheat	—	280	296
Fertilizer Application	2	16		Urea	50 kg.	124	
				M. Potash	50 kg.	60	
				Phosphate	50 kg.	110	310
Cultural Practices	1	8					8
Top Dressing of Fertilizer	1	8		Urea	25 kg.	62	70
Irrigation	5	40		Tubewell	30 hrs.	300	300

TABLE 7.4(A) (*Contd.*)

Harvesting	10	80				80
Threshing	6	54	Machine	5 hrs.	50	104
Total Expenditure						1260

**Draught* : Shows hire charges for a pair of bullocks and plough paid @ Rs. 2 per hour.

TABLE 7.4(B) : **Output per acre in Rabi Season from the Reclaimed Usar Land, Village Ghazipur**

Crop	Yield (in qtls.)		Prevailing Prices (in Rs. per qtl.)		Output (in Rs.)		Total output (in Rs.)	Profit (in Rs.)
	Grain	Straw	Grain	Straw	Grain	Straw		
Wheat	13	14	160	30	2080	280	2360	1100

TABLE 7.5(A) : **Expenditure involved in one acre of Land Reclamation, First Season (Kharif), Crop Paddy, Village Piparpur, Tahsil Amethi. Banjar Land**

Operation	Human Labour (in man-days)	Wages paid Rs.	Draught* Employed and payment made Rs.	Material used			Total Expenditure Rs.
				Name	Qty.	Cost Rs.	
Levelling and Bunding	10	80					80
Land Preparation	5	40	50				90
Gypsum application	2	16		Gypsum	20 Q.	90	106
Impounding of Water	1	8		Irrigation		30	38
Fertilizer application	1	8		Urea	50 kg.	124	
				Shakti Khad	50 kg.	128	
				M. Potash	30 kg.	40	300
Transplanting of seedling including Nursery cost	15	120		Paddy seedling		250	370
Top Dressing of Fertilizer	1	8		Urea	20 kg.	48	56
Plant Protection	1	8		Znso and Blitox		30	38
Harvesting	12	96					96
Threshing	5	40					40
Total Expenditure							= 1214

**Draught* ; Shows hire charges for a pair of bullocks and plough paid @ Rs. 2 per hour.

paration, gypsum application, impounding of water, fertilizer application, transplanting of seedlings including nursery cost, top dressing of fertilizer, plant protection, harvesting and threshing. It can be seen from Table 7.5 (B) that a farmer's output from the reclaimed banjar land, sowing paddy crop, is Rs 1,725 which includes Rs. 1,625 from grain and Rs. 100 from straw. This amounts to a profit of Rs. 511 in the first (kharif) season of reclamation cycle of banjar land, sowing paddy crop.

In the second (rabi) season of reclamation cycle wheat crop is sown through operation of various processes, viz. land preparation, fertilizer application, seedling cost, cultural practices, top dressing of fertilizer, irrigation, harvesting and threshing. All these processes involve a total expenditure of Rs. 1,130 which is given in Table 7.6 (A). It can be seen from Table 7.6 (B) that there is a total output of Rs. 1,880 which includes Rs. 1,600 from the grain and Rs. 280 from the straw. Thus a farmer makes a profit of Rs. 750 from sowing wheat crop in the second season of the reclamation cycle of banjar land, which means that profit lies in reclaiming in both kharif and rabi seasons.

7.2.4 Old Fallow Land

The reclamation of old fallow land is practised in village Raibigo (tahsil Kadipur) in one acre land and the resulting input-output balance is explained in present exercise. There is a total expenditure of Rs. 1,030 in the first season of the reclamation cycle in which paddy crop is sown and it includes all the reclamation and sowing processes, viz. levelling and bunding, land preparation, fertilizer application, seedling cost including nursery cost, top dressing of fertilizer, irrigation, harvesting and threshing which is given in Table 7.7 (A). It has been recorded from Table 7.7(B) that there is a total output of Rs. 1,970 in reclaiming one acre of old fallow land which includes Rs. 1,820 from the grain and Rs. 150 from the straw, there being a profit of Rs. 940 in reclaiming one acre of old fallow land in the first reclamation cycle sowing paddy crop.

In the second (rabi) season of the reclamation cycle wheat crop is sown in the same reclaimed old fallow land of village

TABLE 7.5(B) : **Output per acre in Kharif Season from the Reclaimed Banjar Land, Village Piparpur**

Crop	Yield (in qtls)		Prevailing Prices (in Rs. per qtl)		Output (in Rs.)		Total Output (Rs.)	Profit (Rs.)
	Grain	Straw	Grain	Straw	Grain	Straw		
Paddy	13	19	125	10	1625	100	1725	511

TABLE 7.6(A) : **Expenditure involved in one acre of Banjar Land Reclamation during Second Season (Rabi), Crop, Wheat, Village Piparpur, Tahsil Amethi**

Operation	Human Labour (in mandays)	Wages paid Rs.	Draught* employed and payment made Rs.	Material used			Total Expen-diture
				Name	Qty.	Cost Rs.	
Land preparation	4	32	80				112
Fertilizer Application	1	8		N.P.K.		190	258
Seedling cost	2	16					198
Cultural Practices	1	8					8
Top Dressing of Fertilizer	1	8		Urea		62	70
Irrigation	5	40		Tubewell	20 Hrs.	200	240
Harvesting	10	80					80
Threshing	8	64		Threshing	10 Hrs.	100	164
Total Expenditure							=1130

**Draught* : Shows hire charges for a pair of bullocks and plough paid @ Rs. 2 per hour.

TABLE 7.6(B) : **Output per acre in Second Season from the Reclaimed Banjar Land, Village Piparpur**

Crop	Yield (in qtl)		Prevailing Prices (in Rs. per qtl)		Output (in Rs.)		Total Output (in Rs.)	Profit (in Rs.)
	Grain	Straw	Grain	Straw	Grain	Straw		
Wheat	10	14	160	20	1600	280	1880	750

TABLE 7.7 (A) : **Old Fallow Land. Expenditure involved during the First Season in Reclaiming one acre land, Crop Paddy, Village Raibigo, Tahsil Kadipur**

Operation	Human Labour (in mandays)	Wages paid Rs.	Draught* employed and payment made Rs.	Material used			Total Expenditure Rs.
				Name	Qnty.	Cost Rs.	
Levelling and Bunding	6	48	100				148
Land Preparation	3	24	50				74
Fertilizer application	1	8		Urea	50 kg.	124	
				Compost		50	
				Shakti Khad	25 kg.	64	246
Seedling cost including				Paddy Seedling			
nursery cost	10	80				180	260
Top Dressing of Fertilizer	1	8		Urea	25 kg.	62	70
Irrigation	2	16		Tubewell	8 Hrs.	80	96

TABLE 7.7(A) (*Contd.*)

Harvesting	12	96	96
Threshing	5	40	40
Total Expenditure			=1030

**Draught* : Shows hire charges for a pair of bullocks and plough paid @ Rs. 2 per hour.

TABLE 7.7 (B) : **Output per acre in the First Season of Reclaimed Old Fallow Land, Village Raibigo**

Crop	Yield (in qtls.)		Prevailing Prices (in Rs. per qtl.)		Output (in Rs.)		Total Output (in Rs.)	Profit (in Rs.)
	Grain	Straw	Grain	Straw	Grain	Straw		
Paddy	14	15	130	10	1820	150	1970	940

Piparpur. It has been recorded from Table 7.8(A) that there is total input of Rs. 1,118 involved in reclaiming and sowing of wheat crop in the same one acre of land which includes all the processes, viz land preparation, fertilizer application, seedling cost, irrigation, harvesting and threshing. The total output achieved from one acre of reclaimed old fallow land in the second season of the reclamation cycle in which wheat crop is sown is Rs. 2,240. This includes Rs. 1,920 from the grain and Rs. 320 from straw (*bhusa*). Thus there is a profit of Rs. 1,092 in the second reclamation cycle of old fallow land.

7.2 5 Fallow Land

The fallow land is reclamable with little effort because such lands are left out of cultivation for more than 1 year but less than 2 years. The fallow land reclamation procedure is practised on one acre of land of village Bhadainya (tahsil Sadar Sultanpur). It is observed from Table 7.9(A) that there is a total expenditure of Rs. 1,192 in reclaiming one acre of fallow land in which paddy crop is sown in the first reclamation cycle. The subsequent total output achieved is Rs. 2,280 which is given in Table 7.9(B) and this includes Rs. 2,080 from the grain and Rs. 200 from straw. Hence there is an overall profit of Rs. 1,088 in the first reclamation cycle of fallow land from one acre of land.

In the second (rabi) season wheat crop is sown and including all the processes of reclamation and sowing the total expenditure incurred is Rs. 1,289 on one acre of fallow land in the same land of village Bhadainya which is clear from Table 7.10(A). It is observed from Table 7.10(B) that there is a total output of Rs. 2,760 achieved in sowing wheat crop in the second reclamation cycle of fallow land which includes Rs. 2,400 from grain and Rs. 360 from straw according to the prevailing prices. There is a profit of Rs. 1,471 in the second (rabi) season of the reclamation cycle of fallow land. Thus there is a profit in both kharif and rabi seasons while the profit is more in the second reclamation cycle. Seeing the reclamation cost of fallow land it is found that fallow land may be very easily reclaimed.

TABLE 7.8 (A) : **Old Fallow Land. Expenditure involved during the Second Season Rabi (wheat crop) on one acre of land reclamation, Village Raibigo, Tahsil Kadipur**

Operation	Human Labour (in mandays)	Wages Paid Rs.	Draught* employed and payment made Rs.	Name	Material used Qty.	Material used Cost Rs.	Total Expenditure Rs.
Land Preparation	10	80	60		-		140
Fertilizer application	2	16		Shakti Khad	50 kg.	128	
				Compost		50	
				Urea	50 kg.	124	238
Seedling cost	1	8		Wheat		220	220
Irrigation	4	32		Tubewell	20 Hrs.	200	232
Harvesting	13	104					104
Threshing	8	64		Thresher	15 Hrs.	150	214
		Totat Expenditure					=1148

**Draught* : Shows hire charges for a pair of bullocks and plough paid @ Rs. 2 per hour.

TABLE 7.8 (B) : **Output per acre in the Second Season of Reclaimed Old Fallow Land, Village Raibigo**

Crop	Yield (in qtls.) Grain	Yield (in qtls.) Straw	Prevailing Prices (in Rs. per qtl.) Grain	Prevailing Prices (in Rs. per qtl.) Straw	Output (in Rs.) Grain	Output (in Rs.) Straw	Total Output (in Rs.)	Profit (in Rs.)
Wheat	12	16	160	20	1920	320	2240	1092

TABLE 7.9 (A) : **Fallow Land. Expenditure involved during the First Season, Kharif (crop paddy), in one acre of land reclamation, Village Bhadainya, Tahsil Sadar Sultanpur**

Operation	Human Labour (in man days)	Wages Paid Rs.	Draught* employed and payment made Rs.	Material used			Total Expenditure Rs.
				Name	Qty.	Cost Rs.	
Land Preparation	8	64	40				104
Fertilizer Application	2	16		Urea	50 kg.	124	
				Shakti Khad	25 kg.	64	
				M. Potash	50 kg.	60	
				Compost		50	314
Impounding of water	1	8		Tubewell	5 Hrs.	50	58
Transplanting of seedling including nursery cost	15	120		Paddy Seedling		280	400
Top Dressing of Fertilizer	1	8		Urea	20 kg.	48	56
Irrigation	2	16		Tubewell	10 Hrs.	100	116

TABLE 7.9(A) (*Contd.*)

Harvesting	10	80	80
Threshing	8	64	64
Total Expenditure			=1192

**Draught* : Shows hire charges for a pair of bullocks and plough paid @ Rs. 2 per hour.

TABLE 7.9 (B) : **Output per acre in the First Season of Reclaimed Fallow Land, Village Bhadainya**

Crop	Yield (in qtls)		Prevailing Prices (in Rs. per qtl)		Output (in Rs.)		Total Output (in Rs.)	Profit (in Rs.)
	Grain	Straw	Grain	Straw	Grain	Straw		
Paddy	16	20	130	10	2080	280	2280	1088

TABLE 7.10(A) : **Fallow Land. Expenditure involved in reclaiming one acre of land, Second Season of the Reclamation Cycle, Rabi (crop wheat), Village Bhadainya, Tahsil Sadar Sultanpur**

Operation	Human Labour (in man days)	Wages Paid Rs.	Draught* employed & Payment made Rs.	Material used			Total Expen-diture Rs.
				Name	Qnty.	Cost Rs.	
Land Preparation	8	40					104
Fertilizer application	2	16		S. Phosphate	25 kg.	55	
				M. Potash	50 kg.	60	
				Urea	50 kg.	124	
				Compost		50	305
Seedling cost	2	16		Wheat		250	266
Irrigation	8	64		Tubewell	20 Hrs.	200	264
Top Dressing of Fertilizer	1	8		Urea	25 kg.	64	70

TABLE 7.10(A) (*Contd.*)

Harvesting	12	96				96
Threshing	8	64	Thresher	12 Hrs.	120	184
Total Expenditure						=1289

**Draught* : Shows hire charges of a pair for bullocks and plough paid @ Rs. 2 per hour.

TABLE 7.10(B) : **Output per acre in the Second Season of Reclaimed Fallow Land, Village Bhadainya**

Crop	Yield (in qtls)		Prevailing Prices (in Rs. per qtl)		Output (in Rs.)		Total Output (in Rs.)	Profit (in Rs.)
	Grain	Straw	Grain	Straw	Grain	Straw		
Wheat	16	18	150	20	2400	360	2760	1471

7.2.6 Kankrili Land

It may be less profitable to reclaim kankrili land adopting the recommended practices as discussed in the earlier chapter following the *chari* and barley crop rotation. In the first season of kankrili land reclamation cycle profit may be normal. It can be mentioned that input-output balance depends very much on the level of management and availability of input for the cultivation. The separation of kankar nodules from the soil is a costly affair owing to upward translocation of soluble salts. The removal of the upper soil also exposes the hard soil which is again problematic for the cultivation. If the trails are made for the separation of this hard kankar pan it will be a more desirable practice for the onward cultivation.

The reclamation of kankrili land is practised with in one acre of land in village Seurchamurkha (tahsil Sadar Sultanpur). There is a total expenditure of Rs. 1,028 in the first reclamation cycle (kharif season) in which the fodder crop (*chari*) is sown which requires a low input. This total expenditure includes all the processes of reclamation and sowing of the *chari* crop which is clear from Table 7.11(A). The total output achieved from the first reclamation cycle is Rs. 1,625 which includes Rs. 1,000 from grain and Rs. 625 from straw which is given in Table 7.11(B). Thus there is a profit of Rs. 597 in the first reclamation cycle of kankrili land.

In the second season of reclamation cycle barley crop is sown and including all the processes of reclamation and sowing the total expenditure is Rs. 1,136 in the same reclaimed land as given in Table 7.12(A). It has been recorded from Table 7.12(B) that the total output achieved from the reclaimed one acre of kankrili land in the second season in which barley crop is sown is Rs. 1,905 which includes Rs. 1,680 from grain and Rs. 225 from straw. Thus there is a profit of Rs. 769 in the second reclamation cycle of one acre of kankrili land sowing barley crop. Thus it is observed from the above tables that there is a definite profit in sowing crops after reclamation of kankrili land in both kharif and rabi seasons where crops incurring more expenditure are not sown. It can be recorded from the above discussion that the kankrili land may be reclaimed with great difficulty.

TABLE 7.11(A) : **Kankrili Land. Expenditure involved in reclaiming one acre of land during the First Reclamation Cycle, Kharif (crop *chari*), Village Seurchamurkha, Tahsil Sadar Sultanpur**

Operation	Human Labour (in man days)	Wages paid Rs.	Draught* employed & payment made Rs.	Material used			Total Expenditure Rs.
				Name	Qty.	Cost Rs.	
Separation Process	50	400					400
Land Preparation	5	40	80				120
Impounding of water	1	8		Tubewell	5 Hrs.	50	58
Fertilizer application	2	16		Compost		50	
				Urea	50 kg.	124	140
Seedling cost	1	8		Chari		150	158
Harvesting	10	80					80
Threshing	8	64					64
Total Expenditure							=1028

**Draught* : Shows hire charges for a pair of bullocks and plough paid @ Rs. 2 per hour.

TABLE 7.11(B) : **Output per acre in the First Season of Reclaimed Kankrili Land, Village Seurchamurkha**

Crop	Yield (in qntls)		Prevailing Prices (in Rs. per qntl)		Output in (Rs.)		Total Output (in Rs.)	Profit (in Rs.)
	Grain	Straw	Grain	Straw	Grain	Straw		
Chari	10	125	100	5	1000	625	1625	597

TABLE 7.12(A) : **Kankrili Land. Expenditure involved in reclaiming one acre of land during the Second Reclamation Cycle, Rabi Season, Crop Barley, Village Seurchmurkha, Tahsil Sadar Sultanpur**

Operation	Human Labour (in man days)	Wages paid Rs.	Draught* employed and payment made Rs.	Material used			Total Expenditure Rs.
				Name	Qty.	Cost Rs.	
Re-separation Process	5	40					40
Land Preparation	10	80	80				160
Fertilizer application	1	8		Urea	50 kg.	124	
				Compost		50	182
Seedling Cost	2	16		Barley		180	196
Top Dressing of Fertilizer	1	8		Urea	25 kg.	62	70
Irrigation	3	24		Tubewell	20 Hrs.	200	224
Harvesting	8	64					64

TABLE 7.12(A) (*Contd.*)

Threshing	10	80	Thresher	12 Hrs.	200
	Total Expenditure				=1136

**Draught* : Shows hire charges for a pair of bullocks and plough paid @ Rs. 2 per hour.

TABLE 7.12(B) : Output per acre in the Second Season of Reclaimed Kankrili Land, Village Seurchamurkha

Crop	Yield (in qtls)		Prevailing Prices (in Rs. per qtl)		Output (in Rs.)		Totol Output (in Rs.)	Profit (in Rs.)
	Grain	Straw	Grain	Straw	Grain	Straw		
Barley	14	15	120	15	1680	225	1905	769

7.2.7 Ravine Land

The reclamation cost of ravine land is mainly dependent on the depth and width of ravine. The Government of Uttar Pradesh undertook to reclaim a big area under ravines at Rehmankhera under the State Soil Conservation Programme. Most of the land in the field had been so badly damaged by erosion forming ravines that agricultural practices were impossible. As a result of the adoption of soil conservation and good farm management practices on the arable part of the farm, the land has been completely reclaimed. It is now possible to raise all kinds of crops.

In cooperation with the Kamala Nehru Agriculture Farm Science Centre, Sultanpur, ravine reclamation project is undertaken on one acre ravine land of village Ratanpur (tahsil Sadar Sultanpur). In the reclamation measures all manual and mechanical labour is utilised. Under the mechanical measures tractor is used for levelling the land and manual labour is used for bunding and other cropping processes. It has been recorded from Table 7.13(A) that there is a total expenditure of Rs. 1,068 in reclaiming and sowing one acre of ravine land with maize fodder crop and this total input includes all expenditure, viz. mechanical and manual from the time of reclamation to harvesting the crop. The total output acheived from the cultivated field is Rs. 2,950 which includes Rs. 450 from grain and Rs. 2,500 from straw (fodder) which is given in Table 7.13(B). Thus there is a profit of Rs. 882 on the first reclamation cycle of ravine land in village Ratanpur.

In the second (rabi) season of the reclamation cycle barseem fodder crop is sown and the total expenditure incurred is Rs. 1,456 which includes all the involved processes, viz. levelling and bunding, land preparation, fertilizer application seedling cost, harvesting and threshing as given in Table 7.14(A). It can be seen from Table 7.14(B) that there is a total output of Rs. 2,850 in the second reclamation cycle of ravine land which includes Rs. 600 from grain and Rs. 2,250 from straw (fodder). The total profit achieved in the second reclamation cycle of ravine land is Rs. 1,254. It has been recorded from the above discussion that there is much profit in the first reclamation cycle but the profit has increased tremendously in

TABLE 7.13(A) : **Ravine Land. Expenditure involved in reclaiming one acre of land during the First Reclamation season, Kharif (crop maize fodder), Village Ratanpur, Tahsil Sadar Sultanpur**

Operation	Human Labour (in man days)	Wages Paid Rs.	Draught* employed & payment made Rs.	Material used			Total Expen-diture Rs.
				Name	Qty.	Cost Rs.	
Levelling & Bunding	15	120	1000				1120
Land Preparation	5	40	80				120
Fertilizer application	1	8		Urea	50 kg.	124	
				Shakti Khad	25 kg.	64	
				Compost		50	246
Seedling Cost	1	8		Maize	50 kg.	150	158
Irrigation	5	40		Tubewell	20 kg.	200	240

Harvesting	15	120	120
Threshing	8	64	64
	Total Expenditure		=2068

**Draught* : Shows hire charges for a pair of bullocks and plough paid @ Rs. 2 per hour.

TABLE 7.13(B) **Output per acre in the First Season of Reclaimed Ravine Land, Village Ratanpur**

Crop	Yield (in qtls)		Prevailing Prices (in Rs. per qtl)		Output (in Rs.)		Tolal Output (in Rs.)	Profit (in Rs.)
	Grain	Straw	Grain	Straw	Grain	Straw		
Maize	3	500	150	5	450	2500	2950	882

TABLE 7.14(A) : **Ravine Land. Expenditure involved in reclaiming one acre of land during the Second Reclamation Season, Rabi (crop Barseem fodder), Village Ratanpur, Tahsil Sadar Sultanpur**

Operation	Human Labour (in man days)	Wages Paid Rs.	Draught* employed and Payment made Rs.	Material used			Total Expen-diture Rs.
				Name	Qty.	Cost Rs.	
Levelling and Bunding	8	64	20				84
Land Preparation	4	32	50				82
Fertilizer application	2	16		Urea	50 kg.	124	
				Shakti Khad	50 kg.	128	
				M. Potash	30 kg.	40	
				Compost		50	358
Seedling Cost	1	8		Barseem	20 kg.	500	508
Irrigation	8	64		Tubewell	30 Hrs.	300	364

Harvesting	20	160	160
Threshing	5	40	40
Total Expenditure			1596

**Draught* : Shows hire charges for a pair of bullocks and plough paid @ Rs. 2 per hour.

TABLE 7.14(B) **Output per acre in the Second Season of Reclaimed Ravine Land, Village Ratanpur**

Crop	Yield		Prevailing Prices		Output (in Rs.)		Total Output (in Rs.)	Profit (in Rs.)
	Grain (kg.)	Straw (qtl)	Grain in Rs. per kg.	Straw in Rs. per qtl	Grain	Straw		
Barseem	60	450	10	5	600	2250	2850	1254

the second season of the reclamation cycle of ravine land because the input-output balance mainly depends upon the management and type of crop which is sown and also the nature of market of that particular crop In this case, management of reclamation is more scientific and cropping is also done very carefully.

7.2.8 Other Types of Waste Land

The reclamation measure is practised in one acre of other types of waste land in village Dadara (tahsil Musafirkhana). In the first (kharif season) reclamation cycle jowar crop is sown. The total input incurred in reclaiming and cropping of the jowar crop is Rs. 1,160 which includes all the processes of reclamation and sowing of the crop as given in Table 7.15(A). The total output which is achieved from this one acre of reclaimed other types of waste land in the first season is Rs. 1,760 which includes Rs. 160 from grain and Rs. 1,600 from straw which is given in Table 7.15(B). The total profit made from this one acre of reclaimed other types of waste land is Rs. 600.

In the second (rabi season) reclamation cycle barseem crop is sown in which the total expenditure incurred is Rs. 1,456 which includes all the involved processes, viz. levelling and bunding, land preparation, fertilizer application, irrigation, seedling cost, harvesting and threshing as given in Table 7.16(A). The total output achieved from the same acre of other types of waste land, in the second reclamation cycle in which barseem fodder crop is sown, is Rs. 2,500 which includes Rs. 500 from grain and Rs. 2,000 from straw (fodder) which is given in Table 7.16(B). There is a profit of Rs. 1,044 in the second reclamation cycle of other types of waste land. It is observed from the above discussion that there is a profit in reclaiming and cropping other types of waste land in both kharif and rabi seasons.

According to cost-benefit analysis fallow land followed by other types of waste land, old fallow land, ravine land, usar land, waterlogged land, banjar land and kankrili land are profitable to reclaim. But seeing the cost incurred on reclamation processes of various waste land types it is observed that fallow and other types of waste land are very easily reclaimable, old

fallow land is easily reclaimable, waterlogged land and banjar land are reclaimable with certain difficulty, usar land and ravine land are reclaimable with moderate difficulty and kankrili land is reclaimable with great difficulty. Here it cannot be generalised that the limitation of the waste lands, i.e. very easily reclaimable waste lands should provide more benefit and waste lands with higher limitations of reclamation should provide low profit. The nature of reclamation may be generalised according to the nature and severity of waste land types but the reclamation measures, cropping pattern and the involved input depends upon the types of crop that are sown, proper management, total achieved output and the nature of market for that produced crop. But it has been observed from all the waste land types reclamation that when a crop is sown, it provides much profit as explained earlier but it varies from one waste land type to another, viz. from fallow land in both kharif and rabi seasons maximum profit has been achieved whereas low profit is achieved from the banjar land in both kharif and rabi seasons.

7.3 SUGGESTION FOR EFFECTIVE MANAGEMENT OF WASTE LANDS

A few suggestions for the effective utilisation of waste lands in Sultanpur District are listed below:

Effective measures to check the initiation of waste land is to be undertaken both at the individual and collective levels.

A detailed soil survey of the land use should be carried out in each area proposed for the reclamation of any of the waste land types.

After reclamation measures are carried out, the responsibility of follow up cultivation should be that of the normal staff of the Agricultural Department. The technical staff should suggest suitable cropping patterns and also introduce improved agricultural practices in the reclaimed areas.

The extension staff of the community development organisation can assist particularly in the demonstration of improved methods of cultivation and in arranging for supply of agricultural requisites.

Location of the affected areas of all waste land types according to their development should be made available for the reclamation project.

A Committee should be set up for preparation of waste lands reclamation schemes.

Proposals should be invited for the improvement of all waste lands and advances made for this purpose in accordance with the schedule approved by the Government.

Agricultural cooperative societies be set up to provide all technical assistance and training to the farmers for the reclamation of waste lands.

Facilities for augmenting irrigation in the reclaimed areas and arrangement for loans for the purpose of mechanical inputs should be made available.

Proper follow-up action be ensured for cultivation in the reclaimed lands with a view to improve their productivity.

Allotment of all waste land types to individuals (landless farmers) and to cooperative societies be made.

Pilot projects be started in those areas where the development of waste lands is widely prevalent, so as to demonstrate to the cultivators practical technical methods for reclaiming waste lands.

Special attention be paid to the aspect of settlement work because the selection of settlers is the most important factor contributing to its success, so that the right type of settlers with an aptitude for farming and willingness to perform hard work are selected.

To make the settlement work successful it is necessary that the Block Agency should be associated with it so that the problems of the farmers are attended to as expeditiously as possible. The personal interest of the District Collector, Paragana Officer, Agriculture Officer and Block Development Officer are necessary.

7.4 CONCLUSION

Thus in the initial stages of the reclamation of waste lands the suitable crop according to the nature and type of waste lands must be followed. As a result of the adoption of the soil conservation and good farm management practices on the arable part of the field, the land can be completely reclaimed. On the whole reclamation of all waste land types is beneficial economically. Thus the main factors influencing the economics of reclamation are cost of reclaiming these lands and the anticipated benefits on account of reclamation in terms of production. According to the investigations of waste land reclamation, experiments show favourable returns in crop production. The total output gives a favourable net income according to the type and nature of waste lands and the type of sown crop in the first and second reclamation cycles. Moreover these waste land types may be turned usable and may be made fertile for all time. In all cases, the return is expected to exceed the investment in the first year in both kharif and rabi seasons. And any attempt at reclaiming any waste land type is sure to be successful in terms of input-output balance. So such projects must be taken up and should be efficiently worked in Sultanpur District and in other areas of the country to increase the overall agricultural productivity.

8

CONCLUSIONS

Indian economy is still predominantly agricultural as this sector accomodates 66.52 per cent of the country's working population. Development of agriculture, therefore, is a prerequisite to the overall economic development of the country. It is imperative to extend the area under cultivation to the every possible extent in order to increase the total quantum of food production. In the Indian context, where the pressure on land is already very high, the only hope of extending the net sown area lies in bringing improvements in two land use categories i.e. fallow land and culturable waste land. The crucial questions in this venture area are: (i) the physical characteristics of these categories and an assessment of the possibilities of their management and (ii) the cost-benefit considerations of reclamation of these categories.

Keeping these questions in view, the present study has been attempted to fulfil the following objectives:

(i) Identifying the typology of waste lands at all levels of the district in the spatial and temporal extent;
(ii) Identifying the determinants of waste land formation;
(iii) Establishing the capability classes of different waste lands;
(iv) Analysing the genesis of waste lands and recommending remedies for their reclamation;
(v) Estimating the cost-benefit of reclaiming different categories of waste lands.

In Sultanpur District, the waste lands are classified into eight types, viz. waterlogged land, usar land, banjar land, old fallow land, fallow land, kankrili land, ravine land and other types of waste land. The percentage of area under waste land to total area of the district was recorded to be 12.85% in 1977-78 which has increased to 13.72% in 1981-82. The proportion of waste lands in different categories varies between 4.18% (under banjar land) and 0.1% (under kankrili land). The percentage change in total waste lands over a five-year period has been recorded to be 10.93%. From the viewpoint of individual categories of waste lands the highest change has been recorded in waterlogged land (29.96%) while minimum change has taken place in fallow land (7.29%). There are large variations in the distribution of waste lands within the district at tahsil, block and village level, both in 1977-78 and 1981-82. The size of waste land has increased during these five years excepting that of the waterlogged land in Gauriganj Block where it has decreased (—30.76%). It is observed that about 98% villages of Sultanpur District have waste lands but the extent varies and about 52% of the villages have only10% of their area under waste land.

An analysis of the variables which have generally been responsible for the variations in the extent of waste lands in Sultanpur District has been attempted quantitatively. The following table presents the results of stepwise regression exercise which brings out the variables explaining percentage variations of area under different categories of land to the total area of the village.

Types of waste land	Important variable	Other contributory variables
Waterlogged land	Slope	Drainage density, Behaviour of water table and distance of the canal from the field
Usar land	pH value	Thickness of the top soil, drain-

	of the soil	age density and quality of water in pH.
Banjar land	pH value of the soil	Land concentration, slope and quality of water in pH.
Old Fallow Land	pH value of the soil	Land concentration and behaviour of water table.
Fallow land	Land concentration	Drainage density and pH value of the soil.
Kankrili land	pH value of the soil	Drainage density, behaviour of water table and depth of base rock.
Ravine Land	Drainage density	Slope, behaviour of water table, quality of water in pH and pH value of the soil.
Other types of waste land	Land concentration	Quality of water in pH, drainage density and pH value of the soil.

The genesis of each type of waste lands has been analysed by taking growth in an overall area as an index for the development of waste lands. The larger extent of an area under waste land has been taken as a proxy to its intensity. Thus, the intensity of waste lands has been analysed in three selected villages, having varying intensities of waste land development. The genesis of different types of waste land has been analysed in different stages of waste land development. The table below presents the genetic characteristics of the waste land categories.

Types of waste land	Initial stage	Intermediate stage	Severity stage
Waterlogged land	Limited percolation	Seepage	Surface Accumulation
Usar land	Surface wash	Salt encrustation	Usarisation
Fallow, old fallow and Banjar land	Structural deterioration of the soil	Soil wash	Barrenness
Kankrili land	Slope wash	Stratification of calcium carbonate nodules	Intensive calcification of kankar nodules
Ravine land	Rain wash	Entrenching of channels	Severe tank erosion
Other types of waste land	—	Slope wash	—

The waste land capability classification is a method of grouping waste lands on the basis of their classes assessed in relation to their suitability of agricultural uses. The following table contains the capability classes in different waste land types.

Types of waste land	Capability classification
Waterlogged land	Reclaimable with certain difficulty
Usar land	Reclaimable with moderate difficulty
Banjar land	Reclaimable with certain difficulty
Old fallow land	Easily reclaimable
Fallow land	Very easily reclaimable
Kankrili land	Reclaimable with great difficulty
Ravine land	Reclaimable with moderate difficulty
Other types of waste land	Very easily reclaimable

The methods of reclaiming waste land have been analysed separately for each waste land type. For this purpose eight villages, for each type of watse land, have been selected on the basis of the highest percentage of a particular waste land type occurring in a particular village.

From these sample villages one acre of the affected land has been taken up for reclamation to find out the efficacy of a particular reclamation method. The study has been attempted for both the cropping seasons, i.e. kharif and rabi. The following methods have been tried in some of the sample villages.

(i) In village-Malapur-Jagdishpur improvement of drainage has been adopted as a method to reclaim water logged area. A drain has been dug and connected with another natural outlet for draining out the accumulated water.

(ii) To reclaim usar land in village Ghazipur, mechanical leaching and application of chemicals have been adopted.

(iii) The reclamation of fallow, old fallow and banjar land has been attempted through levelling and bunding, providing drains along the bunds, applying chemicals, after testing the soil and through green manuring.

(iv) For the reclamation of kankar land, compact clay/kankar pan in soil is broken by deep ploughing. The kankar is collected and removed from the field. Proper levelling, bunding is done and chemical fertilizers are used to improve the fertility status of the soil.

(v) The reclamation of ravine land is attempted through levelling it. Strong bunds along the drains are constructed to check the rate of flow of water and dams are also constructed to reduce the erosion.

(vi) The other types of waste land are reclaimed through proper levelling and bunding of the land. Soil erosion has been controlled through dam construction and providing outlets for the excessive run-off. The lands are closed for grazing and the animals are kept away.

It has been found from the experience of reclamation in these villages that the reclamation measures may be suitably modified to meet the specific need of the soil. A reclamation project of any waste land type can be an economic proposition if planned on scientific lines. The profit from reclamation of one acre each of different types of waste land is presented in the table below.

Types of waste land	Achieved profit in Rs.	
	First (Kharif) Season	Second (rabi) Season
Water logged land	655	1075
Usar land	527	1100
Banjar Land	511	752
Old fallow land	940	1092
Fallow land	1088	1471
Kankrili land	597	769
Ravine land	882	1254
Other types of waste land	600	1044

The cost of reclamation varies from one waste land type to another. The cropping and benefit from any crop on a piece of reclaimed land depend upon the inputs, type of crop and agricultural management. Generally the total output has increased from kharif season to rabi season. It is recorded that the highest profit is made from the fallow land reclamation in rabi season. The same pattern is obtained in old fallow land, ravine land, usar land, waterlogged land, other types of waste land and kankrili land. Thus it is found that all the reclaimed waste land types are profitable to the farmers and these lands may be brought under cultivation. It has also been observed that in all cases return is expected to exceed the investment promising a margin of profit in both the crop seasons.

The study offers the following suggestions for managing the waste lands in general.

- — Location of the affected areas of all waste land types be identified before launching the reclamation project.
- — Pilot projects be started in those areas where the development of waste lands is widely prevalent so as to have demonstrable effect on cultivators.
- — Vegetation and engineering controls may be tried to check the ravine formation.
- — After reclamation measures are carried out, the responsibility of follow up action be assigned to the cultivators on the technical advice of the staff of the Agriculture Department. The technical staff should be able to suggest suitable cropping patterns and improved agricultural practices in the reclaimed areas.
- — Facilities for augmenting irrigation in the reclaimed areas and arrangement for loans, for the purpose of mechanical inputs should be made available.

It may not be possible to reclaim every bit of waste lands in the whole expanse of the area under study. But, having carefully studied the cost benefit situation of the reclaimed areas of various waste land types of Sultanpur district, it may be suggested that for increasing agricultural production, waste land reclamation should be adopted as a strategy for the extension of net sown area and increasing the total agricultural production.

APPENDICES

APPENDIX 1

Waterlogged land – Sample Villages, 1981-82

(Area in acres)

Sample villages	Total Area	Water-logged land	Usar land	Banjar land	Old fallow land	Fallow land	Kank-rili land	Ravine land	Other types of waste land	Total waste lands
1	2	3	4	5	6	7	8	9	10	11
Malapur Jagdishpur	1358	298 (21.94)		18 (1.33)	8 (0.59)	9 (0.66)			12 (0.88)	345 (25.44)
Asrawan	82	18 (20.05)							2 (2.10)	20 (22.15)
Katawana	301	59 (19.25)		2 (0.22)					6 (1.99)	87 (21.46)

APPENDIX I (*Contd.*)

1	2	3	4	5	6	7	8	9	10	11
Kamangirpur	35	6	4						1	11
		(17.14)	(11.43)						(2.86)	(31.43)
Kamapur	161	26		3	17	3			3	52
		(16.15)		(1.86)	(10.56)	(1.86)			(1.86)	(32.30)
Hariharpur	205	32		7					8	47
		(15.61)		(3.41)					(3.90)	(22.93)
Nathupur	117	17	13	1					23	54
		(14.53)	(11.11)	(0.85)					(19.66)	(46.15)
Baijipur	159	20	6		7				26	59
		(12.58)	(3.77)		(4.40)				(16.35)	(37.11)
Bokhrepur	457	47	32	55		15	15		85	249
		(10.28)	(7.00)	(12.00)		(3.28)	(3.28)		(18.60)	(54.49)
Tajuddinpur	360	31			20				15	66
		(8.61)			(5.56)				(4.17)	(18.33)
Ugharpur	525	31		28		3			28	90
		(5.90)		(5.33)		(0.57)			(5.33)	(17.14)

Maharani	780	35		21	30			7	93
Pachhim		(4.49)		(2.69)	(3.85)			(0.90)	(11.92)
Mustafabad	1188	38	1	7				40	86
Saraiya		(3.20)	(0.08)	(0.59)				(3.37)	(7.24)
Kadipur	992	13		16	110			3	142
Khurd		(1.31)		(1.61)	(11.09)			(0.30)	(14.31)
Haripur	404	1		8				7	16
		(0.25)		(1.98)				(1.73)	(3.96)

Note : The figures in parentheses represent proportion to total area.

APPENDIX 2

Usar land—sample Villages, 1981-82

(Area in acres)

Sample villages	Total area	Water logged land	Usar land	Banjar land	Old fallow land	Fallow land	Kankrili land	Ravine land	Other Types of waste land	Total waste lands
1	2	3	4	5	6	7	8	9	10	11
Ghazipur	498		199 (39.96)			10 (2.01)			5 (1.00)	214 (42.97)
Purainachandrabhan	161		61 (37.89)	2 (1.24)	3 (1.86)	5 (3.11)				71 (44.10)
Dilipgarh	21		7 (33.33)							7 (33.33)
Bhnaisada	406		126 (32.18)	1 (0.25)		30 (6.98)			3 (0.74)	163 (40.15)

Bhojpur	783		203 (26.93)	8 (1.02)	5 (0.64)			216 (27.59)
Kashipur	42		9 (21.43)					9 (21.43)
Malakdoshipatti	521		93 (17.85)	12 (2.30)				105 (20.15)
Tiwaripur	77		11 (14.29)	6 (7.79)	2 (2.60)	11 (14.29)	1 (1·30)	31 (40.27)
Bhain	745		91 (12.21)	1 (0.13)				92 (12.35)
Purkhipur	243		24 (9.88)	13 (5.35)		6 (2.46)		43 (17.70)
Sherpur	112		8 (7.14)	2 (1.79)	1 (0.89)	1 (0.89)	1 (0.89)	13 (11.61)
Nonara	406		23 (6.16)	9 (2.22)		7 (1.72)	2 (0.49)	43 (10.59)
Kharsoma	333	15 (4.50)	10 (3.00)	20 (6.00)		20 (6.00)	10 (3.00)	75 (22.50)

APPENDIX 2 (*Contd.*)

1	2	3	4	5	6	7	8	9	10	11
Mirdapur		254		4	4					8
				(1.57)	(1.57)					(3.14)
Baithu		542		4		82	82			168
				(0.74)		(15.13)	(15.13)			(31.00)

Note : The figures in parentheses represent proportion to total area.

APPENDIX 3

Banjar Land—Sample Villages, 1981-82

(*Areas in acres*)

Sample villages	Total area	Water logged land	Usar land	Banjar land	Old fallow land	Fallow land	Kankirli land	Ravine land	Other types of waste land	Total waste land
1	2	3	4	5	6	7	8	9	10	11
Piparpur	2519			966	45	47				1058
				(38.34)	(1.79)	(1.87)				(42.00)
Gulalpur	105			40		2			2	44
				(38.10)		(1.90)			(1.90)	(41.90)
Kushela	282	6		96	8	73			7	190
		(2.13)		(34.04)	(2.84)	(25.83)			(2.48)	(67.38)
Mugrahi	65			21	18	7				46

APPENDIX 3 (*Contd.*)

1	2	3	4	5	6	7	8	9	10	11
				(32.69)	(27.69)	(10.38)				(70.77)
Gharwaspur	629			207					46	253
				(32.01)					(7.31)	(40.22)
Baksara	262			76		30				106
				(29.01)		(11.45)				(40.46)
Balpur	382			93					1	94
				(24.35)					(0.26)	(24.61)
Dhabawa	713			157	72	60	15	25	13	342
				(22.02)	(10.10)	(8.42)	(2.10)	(3.51)	(1.88)	(47.97)
Majhui	65			13						25
				(20.00)						(37.46)
Singhauli	486			78					7	85
				(16.05)					(1.44)	(17.49)
Banbhukar	290			43		26			30	99

			(14.83)		(8.97)	(10.34)	(34.14)
Bhagirathpatti	46		4	2		7	13
			(8.70)	(4.35)		(15.22)	(28.26)
Baini	334		18	18		11	47
			(5.39)	(5.39)		(3.29)	(14.07)
Chaukia	575		18	19		7	44
			(3.50)	(3.69)		(1.36)	(8.54)
Ghoraha	1018	6	18		17	3	44
		(0.59)	(1.77)		(1.67)	(0.29)	(4.32)

Note : The figures in parentheses represent proportion to total area.

APPENDIX 4

Old Fallow Land—Sample Villages, 1981-82

(Area in acres)

Sample villages	Total area	Water logged land	Usar land	Banjar land	Old fallow land	Fallow land	Kankrili land	Ravine land	Other Types of waste land	Total waste lands
1	2	3	4	5	6	7	8	9	10	11
Raibigo	2222			65	989	63	13	43	37	1210
				(2.99)	(44.51)	(2.84)	(0.59)	(1.94)	(1.67)	(54.46)
Bhojpur	101				44					44
					(43.56)					(43.56)
Bhagupur	105			1	44	12				57
				(0.95)	(41.90)	(11.43)				(54.29)

Raghaipur	252			99		46	145
				(39.28)		(18.28)	(57.54)
Richhaura	1214			420			420
				(34.60)			(34.60)
Mahmudpur	306	14	13	99	72		198
		(4.58)	(4.25)	(32.35)	(23.53)		(64.71)
Mujesh	289		25	82	28		135
			(8.67)	(28.37)	(9.69)		(46.71)
Dharmdashpur	22			5			5
				(22.73)			(22.73)
Sakardepur	252	3		50	43	8	104
		(1.19)		(19.84)	(17.06)	(3.17)	(41.27)
Saranghpur	942		63	135	36		234
			(6.69)	(14.33)	(3.82)		(24.84)
Sirwara	809	26		82	12	3	123
		(3.21)		(10.14)	(1.48)	(6.37)	(15.20)

APPENDIX 4 (*Contd.*)

1	2	3	4	5	6	7	8	9	10	11
Karondipur	265			10	23	28	10	19	10	100
				(3.77)	(8.68)	(10.57)	(3.77)	(7.17)	(3.77)	(37.74)
Dhundhu	968				52	36			111	199
					(5.73)	(3.92)			(11.47)	(20.56)
Gariynawa	229			5	7				12	24
				(2.18)	(3.06)				(5.24)	(10.48)
Tirahi	163			2	3	3			9	17
				(1.23)	(1.84)	(1.84)			(5.52)	(10.43)

Note : The figures in parentheses represent proportion to total area.

Sherpur	737	21	70	9	41	22	163
		(2.85)	(9.50)	(1.32)	(5.56)	(3.99)	(22.12)
Saidpur	218		8		7	7	20
			(3.67)		(3.21)	(3.21)	(9.17)
Sarwan	950	59	61	5	22	30	177
		(6.21)	(6.42)	(0.53)	(2.32)	(3.16)	(18.63)
Chandpur Sadar	628	48	17		8		73
		(7.64)	(2.71)		(1.18)		(11.62)

Note: The figures in parentheses represent proportion to total area.

APPENDIX 8

Other Types of Waste Land—Sample Villages, 1981-82

(*Area in acres*)

Sample Villages	Total area	Water logged land	Usar land	Banjar land	Old fallow land	Fallow land	Kankrili land	Ravine land	Other types of waste land	Total waste land
1	2	3	4	5	6	7	8	9	10	11
Dadara	280			33	49			49	889	1029
				(1.18)	(1.75)			(1.71)	(32.08)	(36.72)
Kunda Bhairopur	510			2		10			160	172
				(0.39)		(1.96)			(31.37)	(33.73)
Madanpur Devrar	34				1				10	11
					(2.94)				(29.40)	(32.35)

Ranjitpur	260		9				73	82
			(3.46)				(28.00)	(31.54)
Katgharakhurd	239		9	24			62	95
			(3.77)	(10.04)			(25.94)	(39.75)
Rampur Ustada	184		13	16			45	74
			(7.07)	(8.70)			(24.46)	(40.22)
Ramnagar	310		4				67	71
			(1.22)				(21.61)	(22.90)
Chaubepur	163		10		2	4	34	50
			(6.13)		(1.23)	(2.45)	(20.86)	(30.67)
Amkol	723			28	109	8	128	273
				((3.87)	(15.08)	(1.11)	(17.70)	(37.36)
Kunwarpur—1	139	12					22	134
		(8.63)					(15.83)	(24.46)
Semari Dargar	87				1	13	10	24
					(1.15)	(14.90)	(11.40)	(27.59)
Ballipur	144				13		12	25
					(9.03)		(8.33)	(17.30

APPENDIX 8 (*Contd.*)

1	2	3	4	5	6	7	8	9	10	11
Saifullaganj	938				20	82	21	60	54	239
					(2.13)	(8.74)	(2.24)	(6.40)	(5.76)	(25.27)
Karondipur	265			10	23	28	10	19	10	100
				(3.77)	(8.68)	(10.57)	(3.77)	(7.17)	(3.77)	(37.74)
Basauthi	463					36	4	9	5	54
						(7.78)	(0.86)	(1.94)	(1.08)	(11.66)

Note: The figures in parentheses represent proportion to total area.

BIBLIOGRAPHY

(I) BOOKS

Ahmad, E., 1970, *Soil Erosion in India*, New Delhi, Asia.

Anderson, J.R., 1971, "A Geography of Agriculture", Durbuque, Jowa.

Bali, J.S., 1972, *Ravine Reclamation Needs for India*, New Delhi.

Bennett, H.H., 1955, *Elements of Soil Conservation*, McGraw Hill.

Bhattacharya, J.P., 1964, *Study of Soil Conservation for Agricultural Land*, New Delhi.

Blandford, H.F., 1889, *The Climates and Weather of India, Ceylon and Burma*, London.

Brownrigg, F.W., 1898, *Settlement Report*, District Sultanpur.

Chaudhari, P., (ed.), 1972, *Readings in Indian Agricultural Development*, George Allen and Unwin Ltd., London.

Chauhan, D.S., 1966, *Studies in Utilisation of Agricultural Land*, Aggarawala Pub., Agra.

Chislom, M., 1962, *Rural Settlement and Land Use*, Hutchinson Lib.

Ganguli, H.N., 1954, *Restoration of Soil Structures with Gypsum 1 : 8*, Bikaner Gypsum Ltd.

Govinda Rajan, S.V. and Gopala Rao, H.G., 1978, *Studies on Soils of India*, New Delhi.

Gregor, H.F., 1970, *Agricultural Geography, Themes in Research*, Prentice Hall.

Hadimani, A.S., 1960, *Studies on the Waste Lands of Delhi State*, M.Sc., Associate Indian Agricultural Research Institute, New Delhi.

Hall, A., 1947, *The Soil*, London.

Hartshorne, R., 1959, *Perspective on the Nature of Geography*, Dutler and Tanner, London.

Johnston, J., 1972, *Economic Methods*, MacGraw Hills.

Kelley, W.P., 1927, *The Climates of Continents*, Oxford.

Kendall, M.G., 1957, *A Course in Multi-Variate Analysis*, Charles Griffin, London.

King Thomson, 1953, *Water Miracles of Nature*, N.Y. MacMillan.

Kostrowicky, J., 1979, *Hierarchy of World Types of Agriculture*, Poland.

Kostrowicki J., 1965, *Land Utilisation in East Central Europe, Case Studies*, Geographia Polnica, 5, Warsaw, Poland.

Krishnan, M.S., 1960, *Geology of India and Burma*, Madras.

Leather, J.W., 1914, *Investigations of Usar Lands in the United Provinces*, Government Press, Allahabad.

Mahmood Aslam, 1977, *Statistical Methods in Geographical Studies*, Rajesh Publications, New Delhi.

Medlicott, H.B. and Blanford, W.T., 1879, *Mannual of Geology of India.*

Mitra, Ashok and Mukharjee, Sheikher, 1980, *Population, Food and Land Inequality in India: A Geography of Hunger and in Security*, an ICSSR/JNU/FPF Study, Allied Publishers Pvt. Limited, New Delhi, 1980.

Mohammad, Ali, 1978, *Studies in Agricultural Geography*, Rajesh Publishers, New Delhi.

Mohammad, Ali, 1978, *Dynamics of Agricultural Development in India*, Concept Publishing Company, New Delhi.

Monk House, F.J. and Wilkinson, H.R., 1964, "Maps and Diagram" (2nd edition), University Paperbacks, Methuen and Co. Ltd.

Mooreland, H.H., 1912, *The Agriculture of United Provinces*, Allahabad.

Mooreland, H.H. 1913, *Notes on the Agricultural Conditions and Problems of United Provinces*, Allahabad.

Moore, W.G., 1976, *A Dictionary of Geography*, Penguin Books.

Munshull, R., 1967, *Regional Geography*, Hutchinson Lib.

Nevill, H.R., 1903, *District Gazetteer of the District Sultanpur*, Allahabad.

Nimgam, N.M., 1956, *How to Reclaim Saline and Alkali Lands in the Delhi State*, Thesis Associate, Indian Agricultural Research Institute, New Delhi.

Pilai, V.R. and Panikar, P.G.K., 1965, *Land Reclamation in Kerala*, Asia Publishing House, New Delhi.

Porter, F.W., 1888, *Final Settlement Report of the Sultanpur District*, North Western Provinces and Oudh. Allahabad.

Raheja, P.C., 1966, *Soil Productivity and Crop Growth*, Asia Publishing House, New Delhi.

Rao, K.L., 1979, *India's Water Wealth : It's Assessment, uses and Projections*, Orient Longman.

Raychaudhuri, S.P., 1962, *Soil of India*, New Delhi.

Raychaudhuri, S.P., 1966, *Land and Soil*, National Book Trust, New Delhi.

Robinson, G.M., 1937, *Mother Earth*, London.

Shafi, M., 1960, *Land Utilisation in Eastern Uttar Pradesh*, A.M.U., Aligarh.

Sharma, H.S., 1980, *Ravine Erosion in India*, Concept Publishing Co., New Delhi.

Sharma, P.S., 1973, *Agricultural Regionalisation of India*, New Heights Publishers and Distributors, Delhi.

Shome, K.B. and Raychaudhury, S.P. 1960, *Rating of Soils of India*, Indian Agricultural Research Institute FNI, 28th Sept. 1960, New Delhi.

Siddiqui, Asiya, 1973, *Agrarian Change in the Northern Indian State (U.P. 1819-1933)*, Oxford Clarendon Press.

Singh, Abhalakshmi, 1978, *Economics and Geography of Agricultural Land Reclamation*, B.R. Publishing Corporation, Vivekanand Nagar, Delhi.

Singh, Jasbir, 1974, *An Agricultural Atlas of India : A Geographical Analysis*, Vishal Publications, Kurukshetra (Haryana).

Singh, Jasbir, 1976, *An Agricultural Geography of Haryana*, Vishal.

Singh, M. "Land Utilization in Eastern Uttar Pradesh", Ph. D. Thesis (unpublished).

Singh, N.T., 1972, *Reclamation of Alkali (Kallar) Soils*, Punjabi Agricultural University, Ludhiana.

Singh, R.L., 1971, "India Regional Geography", Varanasi.

Singh, Shrinath, 1976, *Modernization of Agriculture : A Case Study of Eastern Uttar Pradesh*, Heritage Publishers, New Delhi.

Spate, O.H.K. and Learmonth, A.S.J.A., 1968, *India and Pakistan*, London.

Stamp, L.D., 1960, *Applied Geography*, Penguin Books, London.

Stamp, L.D., 1960, *Our Developing World*, London.

Stamp, L.D., 1962, *The Land of Britain, Its Use and Misuse*, London.

Stamp, L.D., 1960, *Our Developing World*, London.

Subramanian, C., 1979, *The New Strategy of Indian Agriculture*, Vikas Publishing House, New Delhi.

Symons, L., 1968, *Agricultural Geography*, Bell's, London.

Wadia, D.N., 1960, *Geology of India*, London.

(II) ARTICLES

Agarwal, L.T. and Mehrotra, C.L., 1951 "Soil Survey and Soil Work in U.P.", Vol. I, Allahabad.

1952 "Soil Survey and Soil Work in U.P.", Vol. II, Allahabad.

1953 "Soil Survey and Soil Work in U.P.", Vol. III, Allahabad.

1954 "Soil Survey and Soil Work in U.P., Vol. IV, Allahabad.

Agarwal, R.R., 1957, "Alkali Soils can be Reclaimed", *Indian Farming* 7(9).

Agarwal, R.R., 1959, "Potential and Economic Aspects of Utilization of Saline Alkali Soils", *Proceedings of Simla Seminar on Recent Advanced in Agronomy and Soil Science*, (*Indian Council of Agricultural Research*, I.A.R.I.).

Agarwal, R.R. and Gupta, C.P., 1957, "Spread and Intensity of Soil, Alkalinity with canal Irrigation in Gangetic Alluvium of U.P.", *Indian Journal of Agricultural Sciences*, 27.

Agarwal, R.R. and Gupta, R.N., 1968, "Saline Alkali Soils in India", *I.C.A.R. Technical Bulletin* (Agricultural Series), No. 15.

Agarwal, R.R. and Mehrotra, C.L., 1966, "Quality of Irrigation Waters in U.P.", *Indian Journal of Agricultural Sciences*, 31.

Agarwal, R.R. and Yadav, J.S.P., 1954, "Saline and Alkali Soils of the Indian Gangetic Alluvium in U.P.", *Journal of Soil Science*, 5(2).

Agarwal, R.R. and Mehrotra, C.L., 1950-58, "Soil Survey and

Raheja, R.C. et al., 1962, "Study of Suitable Trees, Grasses and Crops for Saline and Alkali Land", Prepared by the *Working Group on Waste Land including Saline, Alkali and Water Logged Lands* Government of India.

Raj, S.K.T., 1951, "Ganga Khadir, A study on Land Reclamation in India", *Ministry of Information and Broadcasting* Pamphlet No. 4, New Delhi, p. 44.

Raychaudhuri, S.P., 1971, "Extensions of cultivated area for Increased Agricultural Production", *Proceedings of the Symposium on Science and India's Food Problem*, ICAR Publication.

Raychaudhury, S.P. and Biswas, N.R.D., 1954, "Saline Alkali Soils of Asia with Particular Reference to India, *Transactions of 5th International Congress.*

Raychaudhuri, S.P., 1952, "Final Report of the All India Soil Survey Scheme", *Bulletin Indian Council of Agricultural Research* (13).

Shafi, M., 1968, "The Problems of Waste Lands in India", Published in *The Geographer* Special Number, XXI, International Geographical Congress India, Vol. XV, November 1968, Aligarh Muslim University Geographical Society, Aligarh (India).

Sharma, H.S., 1968, "Genesis of Ravines of the Lower Chambal Valley, India Selected Papers, *21st International Geographical Union Congress*, 1, 114-18.

Signond A.A. Joe, 1924, "The Alkali Soils in Hungary and their Reclamation Measures, in *Soil Science* 18, pp. 379-81.

Srivastava, P.B.L., Mehrotra, G.L. and Agarwal R.R., 1962, "The Effect of Leaching Saline, Alkali Soil with Irrigation waters of different kinds on the Permeability and Composition of the soils and the composition of Leachets", *Journal Indian Society, Social Science*, 10.

Stamp, L.D., 1931, "The Land Utilization Survey of Britain", *Geographical Journal*, London, Vol. 78.

Subhashranjan, Basu, 1980, "Physical Factors Affecting Agriculture : A Geographical Analysis", *Perspectives in Agricultural Geography*, ed. by Noor Mohammad, Concept Publishing Co., New Delhi.

Thomas, K.M. and Srinivasan, A.R., 1949, "Weed Killers", *Indian Farming*, Vol. X, No. 2.

Taylor, C.C., 1948, "Weed Control", *Indian Farming*, Vol. IX, No. 4.

Taylor, C.S., 1909, "Effects of Salt on Soils, *Quarterly Journal, Department of Agriculture*, Bengal.

Taylor, E.M., 1940, Making Land Reclamation Profitable, *Indian Farming*, 1(9).

Uppal, H.L., 1961, Reclaiming Alkali Waste Lands, *ICAR, Bulletin* 85, New Delhi.

Uppal, H.L., 1962, "Reclamation of Saline and Alkali Soils, *Land Reclamation, Irrigation and Power Research Institute*, Punjab.

Uppal, H.L., et al., 1961, "Reclamation of Saline and Alkaline Lands", India, *Directorate of Extension Farm*, No. 66, New Delhi.

Uppal, H.L., Agarwal, R.R. and Kibe, M., 1961, "Alkaline Lands" *Farm Bulletin* 66, *Farming Information Unit* Directorate of Extension, Ministry of Food and Agriculture, New Delhi.

Wadia, D.N. and Auden, J.B., 1939, "Geology Structure of Northern India" *Memoirs of Geological Survey of India*, Vol. LXXIII.

Whyte, R.D. and Sisam, J.W.B., 1949, "The Establishment of Vegetation on Industrial Waste Land", *Common Agr. Bur. Joint*, Pub. No. 14.

Yadav, J.S.P., 1956, "Usar Soils. Their Characteristics and Exploitation in Forestry, *Proc. 9th Science* Congress Part I.

Yadav, J.S.P. and Agarwal, R.R., 1959, "Dynamics of Soil changes in the Reclamation of Saline Alkali Soils of the Indo Gangetic Alluvium", *Journal of Indian Society Soil Science* 7.

Yadav, J.S.P. and Agarwal, R.R., 1961, "A Comparative Study of Effectiveness of Gypsum and Dhaincha in the Reclamation of Saline-Alikali soils", *Journal of Indian Society Soil Science*, 9.

(III) GOVERNMENT AND U.N. PUBLICATIONS

Academy of Sciences of the USSR, 1958, *The application of Drainage in the Reclamation of Salinized Soils*, translated from Russian, USSR, Moscow.

Agricultural Census of Uttar Pradesh, 1970-71, Agricultural Statistics of Uttar Pradesh, 1971.

Bulletin of Agricultural Statistics for Uttar Pradesh (1970-71) Joint Director for Statistics, Directorate of Agriculture, U.P. (India), 1972, p. 16.

Census of India, 1981, Series I—India, Part IIB(i), Primary Census Abstract, General Population Table.

Government of India, *Indian Agriculture in Brief*, Eighth Edition.

District Census Hand Book, Sultanpur District 1971, Series-21, U.P., Part X-B, Primary Census Abstract.

F.A.O., 1976, A Frame work for land evaluation, *Soil Bulletin* No. 1.

I.C.A.R., 1958, Final Report of All India Soil Survey Scheme.

I.C.A.R., 1969, *Hand Book of Agriculture.*

Indian National Science Academy, 1968, "Symposium on Reclamation and use of Waste Lands in India", Proceedings, Bulletin No. 44, INSA, New Delhi.

Ministry of Information and Broadcasting, Government of India, *India 1972 : A Reference Mannual*, the Research and Reference Division, p. 250.

Ministry of Food and Agriculture, Government of India, 1961, Waste Land Survey and Reclamation Committee, *Report on the Location and Utilization of Waste Lands in India*, Part X, U.P., New Delhi.

National Botanic Gardens, Lucknow, *Conquest of "Usar"*, 1961 (An Experiment in Reclamation of Alkali land at Banthara, Lucknow District by Biological and Agronomic Methods), pp. 24.

Planning Commission, Programme Evaluation Organization. 1964, Study of Soil Conservation Programme for Agricultural Land, New Delhi.

Planning Commission, Committee on Natural Resources, 1965, Study of Survey and Reclamation of Ravines in India.

Planning Commission, Committee on Natural Resources, 1963, Study on Waste Lands, Including Saline, Alkali and Water Logged Lands and their Reclamation Measures, New Delhi.

Proceedings of the National Institute of Sciences of India, Vol. 26 (Supplement I), Part A : Physical Sciences 1960, National Institute of Sciences of India, New Delhi.

State Soil Conservation Research 1961, *Annual Report*, State Soil Conservation Research, U.P.

Statistical Book, District Sultanpur, 1981-82, Sultanpur.

Survey of India Topographical Sheets No. 63 F/11, 63 F/10, 63 F/12, 63 F/14, 63 F/15, 63 F/16, 63 J/3, 63 J/4, 63 J/7, 63 J/8, 63 J/11, 63 J/12, 63 K/5, 63 K/9.

The Times of India, Directory and Year Book including Who's Who, 1972, p. 192.

United Provinces of Agra and Oudh Department of Agriculture 1909, Reclamation of Usar Land, Report of Agra and Oudh, Department of Agriculture U.P., 1908-09.

Usar Land Reclamation Committee U.P. 1938, "Report of Usar Land Reclamation Committee", Superintendent, Printing and Stationery, Allahabad.

U.S.A. Salinity Lab. 1954, *Diagnosis and Improvement of Saline and Alkali Soils Handbook*, U.S. Department of Agriculture, No. 60.

U.S.D.A. "A Mannual on Conservation of Soil and Water", *Handbook of Professional Agricultural Workers*, Oxford Pub. Co. 1984.

The Emperial Gazetteer of India, 1974 Vol XXIII, Singhbhumi to Trashi-Chod-Zong, Reprint Edition, Today and Tomorrow Printers and Publishers, 24B/5, Original Road, Karol Bagh, New Delhi, pp. 130-137.

Waste Land Survey and Reclamation Committee, Ministry of Food and Agriculture, 1961, "Report on Location and Utilisation of Waste Lands in India", Part X, Uttar Pradesh.